The Instrumental Spectrometric and Spectroscopic Analysis of Common Medicines

A Handbook

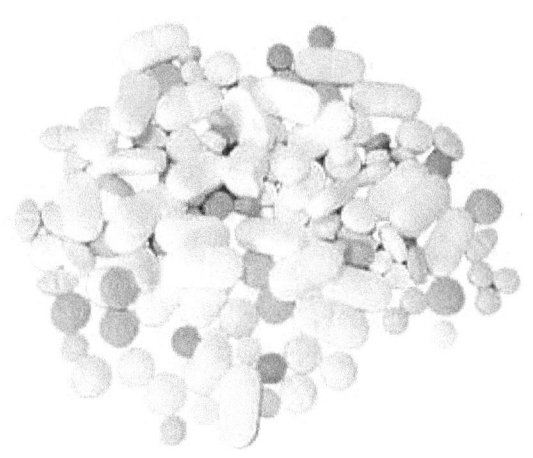

Jasmine Tripconey

and

Nick Winstone-Cooper

First published by Ellixia Publishing Limited in 2023 (www.ellixia.com)

978-1-4717-2892-1

Acknowledgements: Most images and diagrams have been created by the writers and those from other sources are acknowledged with thanks for the permission to use them.

Written for

and the Welsh Government's educational Hwb.

Dr. Nick Winstone-Cooper studied chemistry and physics at Cardiff University before completing postgraduate research in nuclear chemistry, focusing on the creation of radioactive complexes of macrocyclic phosphines for application as heart and bone imaging agents in cancer diagnosis.

He worked extensively in the United Kingdom, France, North America, Italy and the Republic of Korea before moving into education and education consultancy.

Jasmine Tripconey studied Chemistry and Drug Discovery at the University of Bath. Her interests include the biochemistry of selective serotonin reuptake inhibitors and bioinorganic chemistry.

Table of Contents

Introduction

This is the sixth in a series of chemistry handbooks which focus on the instrumental analysis of important chemical substances as an interesting way to develop an ability to interpret the infrared, mass and ^{1}H and ^{13}C nmr spectra of organic compounds. The handbooks are intended to bridge A-level and first year undergraduate chemistry courses and the molecules have been selected carefully to be appropriate to the nature of what students are expected to understand. Too often, students are introduced to chemical compounds which can seem to have been selected at random and so, in these handbooks, the application of the molecules is explained.

Each of the chapters follows the same style with an introductory page discussing the application and history of the development of the medicine with a summary of the elemental data and the formula mass, leading to the determination of the empirical and molecular formulas of the compound. In most cases we have used the trivial/common name for the medicine but for two, aspirin and paracetamol, we have used the commercial namse by which they most commonly known.

The following sections commence with examination and interpretation of the infrared spectrum. Until the introduction of routine multinuclear NMR spectroscopy, infrared (ir) spectroscopy, together with mass spectrometry was one of the most important analytical techniques available. ir spectroscopy is still of fundamental importance as it provides evidence for the presence or absence of particular functional groups but it is important to remember that no ir spectrum can quantify the number of particular functional groups. The spectrum does, however, give clues to the possible structures of the molecule. When there are a number of feasible isomers, however, the ir spectrum rarely allows us to differentiate between them.

Having interpreted the ir spectrum which can give important clues to the structure of the molecule we move on to examining the mass spectrum of the compound. All of the spectra are electron ionisation mass spectra. The absence of fragments predicted from the possible structures means nothing since, even though the travel of the ionised fragments is very brief, they can fragment on their journey. The most important and well known example of this is the McLafferty rearrangement of carbonyl, C = O, containing molecules.

The presence of certain fragments is, however, of profound importance and the most notable example is the peak at m/z = 77 which is only ever due to a benzene ring with one substituent, C_6H_5 –. Similarly, peaks at m/z = 76 which is usually assignable to a benzene ring with two substituents. There are three isomers of disubstituted benzene rings, denoted as ortho – , meta – and para –, and it is rare that the mass spectrum will determine which isomer is present and that is where multi-nuclear nuclear magnetic resonance (NMR) spectroscopy demonstrates its overwhelming significance.

NMR spectra contain three important measurements: the chemical shift, the multiplicity of the peaks and the integrals of the peaks. The ^{13}C nmr spectra only ever contain singlets due to the fact it is only ^{13}C isotopes which resonate and since they are much rarer than ^{12}C isotopes (comprising only about 1.1% of all stable carbon isotopes) there is an extremely low likelihood of two being adjacent in a molecule. If they did occur then there would be coupling and hence the signals would be multiplets. Nevertheless the ^{13}C nmr spectrum is extremely important since the range of chemical shifts is large (δ 0 – 220 ppm) and the regions are clearly differentiated.

This permits conclusions to be drawn and, as we shall see, they are often a better starting point than the ^{1}H nmr spectra.

The ^{1}H nmr spectra contain three important different facets: the chemical shift identifies the type of environment occupied by the hydrogen atoms, the integral informs us of the proportion of all the hydrogen atoms that are chemically and magnetically equivalent whilst the multiplicity indicates the number of hydrogen atoms bonded to adjacent carbon atoms.

Together, the ^{1}H and ^{13}C nmr provide exceptional evidence for the structure of the molecule but this also depends on knowledge of the molecular formula and evidence from the infrared and the mass spectra. If any of the spectra contradicts a proposed structure then the structure must be wrong and it is then time to start again.

Each chapter concludes with display of the concluded structure and the molecule's systematic name.

Part I

Elemental data and correlation charts

The following pages present:-

- A truncated Periodic Table showing the first four periods of the elements; which include all those elements involved in the compounds discussed in this series;

- A correlation chart for the analysis of infra red spectra;

- A concise summary of assignable fragments in the electron ionisation (EI) mass spectrometry analysis. Although there are a variety of ionisation techniques we only consider the simplest and, most commonly used, technique.

- Correlation charts and coupling constants for ^{1}H and ^{13}C nmr spectroscopy;

The Periodic Table of the Elements

Key
atomic number
Symbol
name
relative atomic mass

1	2											(3)	(4)	(5)	(6)	(7)	(0)
(1)	(2)											13	14	15	16	17	18
1 **H** hydrogen 1.0																	2 **He** helium 4.0
3 **Li** lithium 6.9	4 **Be** beryllium 9.0											5 **B** boron 10.8	6 **C** carbon 12.0	7 **N** nitrogen 14.0	8 **O** oxygen 16.0	9 **F** fluorine 19.0	10 **Ne** neon 20.2
11 **Na** sodium 23.0	12 **Mg** magnesium 24.3	3	4	5	6	7	8	9	10	11	12	13 **Al** aluminium 27.0	14 **Si** silicon 28.1	15 **P** phosphorus 31.0	16 **S** sulfur 32.1	17 **Cl** chlorine 35.5	18 **Ar** argon 39.9
19 **K** potassium 39.1	20 **Ca** calcium 40.1	21 **Sc** scandium 45.0	22 **Ti** titanium 47.9	23 **V** vanadium 50.9	24 **Cr** chromium 52.0	25 **Mn** manganese 54.9	26 **Fe** iron 55.8	27 **Co** cobalt 58.9	28 **Ni** nickel 58.7	29 **Cu** copper 63.5	30 **Zn** zinc 65.4	31 **Ga** gallium 69.7	32 **Ge** germanium 72.6	33 **As** arsenic 74.9	34 **Se** selenium 79.0	35 **Br** bromine 79.9	36 **Kr** krypton 83.8

iii

Infrared Correlation Chart

Diagram (absorption regions plotted against wavenumber / cm⁻¹):

Axis: 3600 3400 3200 3000 2800 2600 2400 2200 2000 1800 1600 1400 1200 1000 800 600 400

Labels: RO – H | N – H | =C – H | ArC – H | – C – H | C ≡ N | RCOO – H (carboxylic acids) | C = O | C = C | C = N | Ar – H | C – C | C – O | C – X

Bond	Functional group	Wavenumber (cm⁻¹)
C – C	Alkanes and alkyl groups	750 – 1100
C – X	Haloalkanes (X = Cl, Br or I)	500 – 800
C – F	Fluoroalkanes	1000 – 1350
C – O	Alcohols, carboxylic acids and esters	1000 – 1300
Aliphatic C = C	Alkenes	~ 1650
C = O	Aldehydes, ketones, carboxylic acids, esters and acid chlorides	~ 1750

Bond	Functional group	Wavenumber (cm⁻¹)
Aromatic C = C	Aromatic compounds	1450 – 1650 (Multiple peaks)
C ≡ N	Nitriles	~ 2250
C – H	Alkyl groups, alkenes and aromatic compounds	2850 – 3000 (alkanes) 3000 – 3200 (alkenes and aromatics)
O – H	Carboxylic acids	2500 – 3500
N – H	Amines and amides	3300 – 3500
O – H	Alcohols and phenols	3200 – 3600

Notes: In the table above, Ar refers to an aromatic ring such as benzene whilst X refers to any of F, Cl, Br or I.

Some peaks are of little use for identification purposes since, for example, most organic compounds contain C – H bonds and so the presence of peaks just below 3000 cm⁻¹ is of little use for identification purposes. There is, however, a distinction between the C – H peaks above and below this wavenumber: aromatic compound C – H bonds appear above i.e. to the left of 3000 cm⁻¹ whilst aliphatic C – H bonds appear below, to the right, of 3000 cm⁻¹.

The combination of peaks is also important. For example, a carboxylic acid contains both a C =O and O – H bond so the presence of both is necessary for the confirmation of this class of compound.

Common Mass Fragmentation Ions

The following table shows a large number of fragmentation assignments which are relevant to the molecules in this volume. There are a number of important points to note:-

If a molecule contains a *chlorine* atom then that fragment will exhibit two peaks, two units apart, due to the existence of the ^{35}Cl and ^{37}Cl isotopes. The heights of these peaks will be in the proportion 3:1 and there will always be two other peaks at m/z = 35 and 37, also in the ratio 3:1 due to the natural occurrence of the isotopes. If any of these four peaks are absent then chlorine is not present. The presence of chlorine will already, however, be recorded in the elemental composition of the compound.

Similarly to chlorine, if a molecule contains a *bromine* atom then there will be two peaks for the molecular fragment due to the existence of ^{79}Br and ^{81}Br. Since these two isotopes exist in nearly equal proportions then the peak heights of these fragments will be of approximately equal height and there will, of course, also be peaks of equal height at m/z 79 and 81.

Aromatic compounds such as those containing a benzene ring will usually exhibit a peak at m/z = 77 due to the existence of the C_6H_5 – functional group. If there is a peak at this m/z ratio then it is almost always due to this. There will then also be peaks of lower m/z value due to fragmentation of the ring but these can also be due to the fragmentation of aliphatic chains and so the clue is in the m/z = 77 peak. More highly substituted benzene rings will exhibit peaks at m/z = 76, 75 etc; The existence of the aromatic portion of the molecule is, however, also and *always* conclusively demonstrated by the ^1H and ^{13}C nmr spectra. In the following table, the aromatic fragments are indicated by [a].

Many mass spectrometry measurements may be completed in little more than one second. Whilst in everyday life one second is very brief it is, in physical terms, quite long. This means that the unstable ionised molecule may fragment or rearrange itself on its journey along the apparatus resulting in peaks that would not be predicted simply by considering the ripping apart of a molecule. peaks in the table below which result from rearrangement of an ionised molecule or fragments are indicated by [b] after the fragment's formula.

Some molecules may produce different fragments of the same m/z and are listed as bullet points in the table below.

Table of mass fragments

m/z	Ion	m/z	Ion
15	$[CH_3]^+$	65	$[C_5H_5]^{+a}$
17	$[OH]^+$	67	$[C_5H_7]^+$
18	$[H_2O]^+$	69	$[C_5H_9]^+$
26	$[CN]^+$	70	$[C_5H_{10}]^+$ or $[C_4H_6O]^+$
27	$[C_2H_3]^+$	71	$[C_5H_{11}]^+$ $[C_3H_7\text{-}C\text{=}O]^+$

m/z	Ion	m/z	Ion
28	$[C_2H_4]^+$	72	$[C_2H_5\text{-}CO\text{-}CH_2+H]^{b}$
29	$[C_2H_5]^+$ $[CHO]^+$	73	$[C_3H_7OCH_2]^+$ $[C_3H_7CHOH]^+$ $[C_2H_5O\text{-}C=O]^+$ $[C_2H_5OCHCH_3]^+$
30	$[CH_2NH_2]^+$	74	$[CH_2\text{-}COOCH_3+H]^{+b}$
31	$[CH_2OH]^+$ / $[OCH_3]^+$	75	$[C2H5O\text{-}C=O+2H]^{+b}$ / $[C2H5COO+2H]^{+b}$
35 & 37	$[^{35}Cl]^+$ & $[^{37}Cl]^+$	77	$[C_6H_5]^{+\ a}$
39	$[C_3H_3]^{+\ a}$	78	$[C_6H_5+H]^{ab}$
40	$[CH_2CN]^+$	79	$[C_6H_5+2H]^{ab}$ $[^{79}Br]^+$
41	$[C_3H_5]^+$ / $[CH_2CN+H]^{+b}$	81	$[C_6H_9]^+$ / $[^{81}Br]^+$
42	$[C_3H_6]^+$	82	$[C_6H_{10}]^+$ $[C^{35}Cl^{35}Cl]^+$
43	$[C_3H_7]^+$ $[CH_3C=O]^+$	83	$[C_6H_{11}]^+$ $[CHCl_2]^+$ (also 85&87)
44	$[CH_3CH\text{-}NH_2]^+$	84	$[C_6H_{12}]^+$ $[C^{35}Cl^{37}Cl]^+$
45	$[CH_3CHOH]^+$, $[OCH_2CH_3]^+$, $[CH_2CH_2OH]^+$, $[COOH]^+$	85	$[C_6H_{13}]^+$ $[C_4H_9\text{-}C=O]^+$
49	$[CH_2^{35}Cl]^+$	86	$[C_3H_7\text{-}CO\text{-}CH_2+H]^{+b}$ $[C^{37}Cl^{37}Cl]^+$
50	$[C_4H_2]^{+\ a}$	88	$[CH_2\text{-}COOC_2H_5+H]^{+b}$
51	$[CH_2^{37}Cl]^+$ $[C_4H_3]^{+\ a}$	89	$[C_3H_7\text{-}O\text{-}C=O+2H]^{+b}$ $[C_3H_7COO+2H]^{+b}$
52	$[C_4H_4]^{+\ a}$	90	$[C_6H_5\text{-}CH]^+$
53	$[C_4H_5]^+$	91	$[C_6H_5\text{-}CH_2]^+$ $[C_6H_5\text{-}CH+H]^{+b}$
54	$[H_2CH_2CN]^+$ $[CH_3CHCN]^+$	92	$[C_6H_5\text{-}CH_2+H]^{+b}$
55	$[C_4H_7]^+$	93	$[C_7H_9]^+$ $[CH_2^{79}Br]^+$
56	$[C_4H_8]^+$	94	$[C_6H_5O+H]^+$
57	$[C_4H_9]^+$ $[C_2H_5\text{-}C=O]^+$	95	$[CH_2^{81}Br]^+$
58	$[CH_3\text{-}CO\text{-}CH_2+H]^{b}$	97	$[C_7H_{13}]^+$
59	$[C_2H_5OCH_2]^+$ $[CH_3O\text{-}C=O]^+$ $[C_2H_5CHOH]^+$ $[CH_3O\text{-}CHCH_3]^+$	105	$[C_6H_5C=O]^+$ $[C_6H_5\text{-}CH_2CH_2]^+$
60	$[CH_2\text{-}COOH+H]^{+b}$	107	$[C_6H_5\text{-}CH_2O]^+$
61	$[CH_3COO+2H]^{+b}$ / $[CH_3OCO+2H]^{+b}$	108	$[C_6H_5\text{-}CH_2O+H]^{+b}$
63	$[C_5H_3]^{+a}$	119	$[C_6H_5\text{-}C(CH_3)_2]^+$

a Good diagnostics for benzene ring compounds. *b* Where a fragment results from a rearrangement with the movement of **one** hydrogen atom from one carbon to another this is indicated by the fragment followed by +H.

^{1}H and ^{13}C NMR Correlation Charts

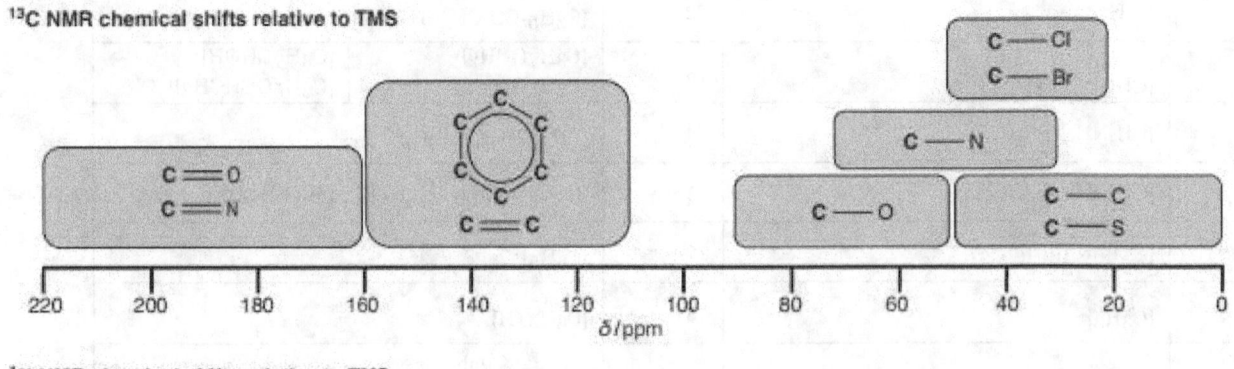

^{13}C NMR chemical shifts relative to TMS

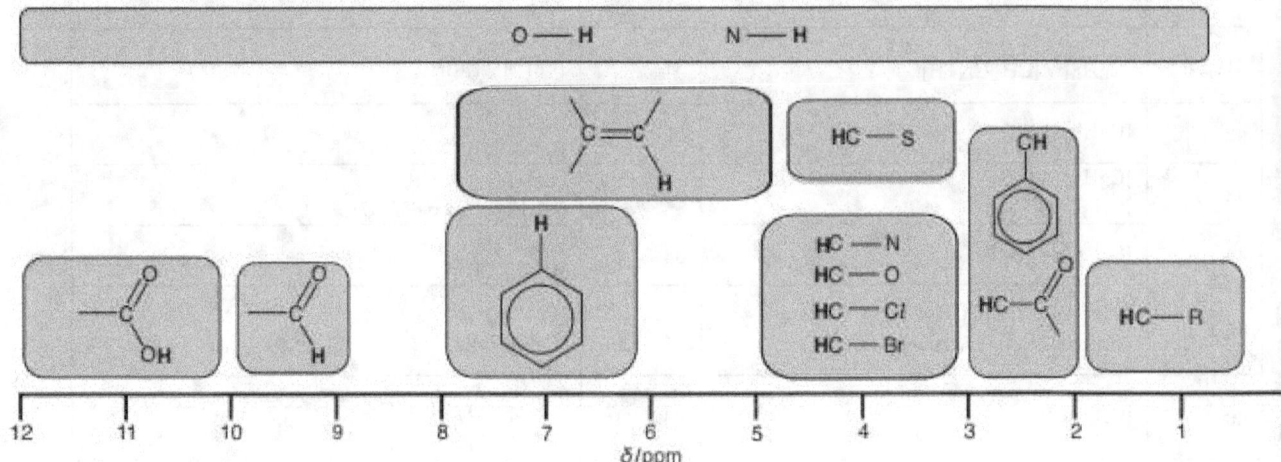

^{1}H NMR chemical shifts relative to TMS

^{1}H NMR Coupling Constants

Aliphatic Alkenes

Isomerism	Coupling constant (J) range (Hz)
Geminal	0 – 5
Vicinal (cis) / Vicinal (Z –)	5 – 14
Vicinal (trans) / Vicinal (E –)	15 – 20

Substituted Aromatic Compounds

Designation	Ortho –	Meta –	Para –
Structural formula			
Coupling constant range (J):	7 – 10	2 – 3	0 – 2

Chapter I

Glutarimide

Glutarimide is a precursor of vast number of many pharmaceuticals including thalidomide and cycloheximide which is a natural fungicide.

Although thalidomide is notorious now for birth defects it is effective as a treatment for leprosy in women beyond child bearing age.

With a melting point of 155°C and a boiling point of 212°C, glutarimide is a crystalline solid at room temperature.

With a formula mass (M_r) of 113.11 g mol^{-1}, glutarimide has the elemental composition:
C: 53.05%, H: 6.25%, N: 12.38%, O: 28.29%

This means that the **empirical** and **molecular** formulas are both $C_5H_7NO_2$.

I – Glutarimide

Infrared Spectrum

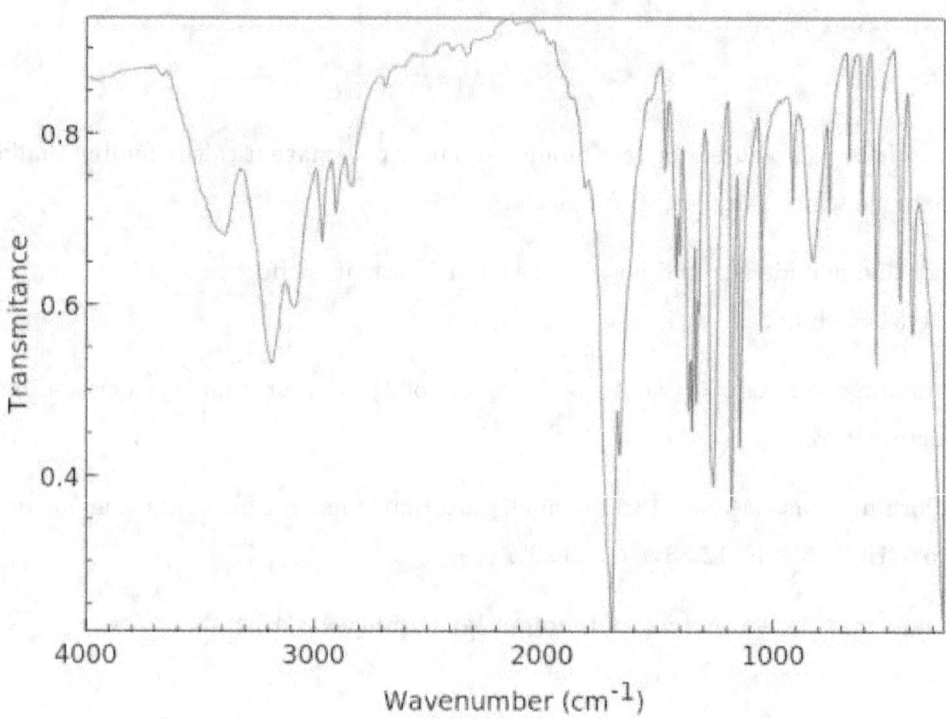

Observations

(√ / X)	Wavenumber range (cm⁻¹)	Wavenumber (cm⁻¹)	Assignment
X	3200 - 3700		O – H
√	3200 - 3600	3460	N – H
?	3000 – 3300	3190, 3060,	C – H (aromatic)
√	2500 – 3000	2920, 2900, 2800	C – H (aliphatic)
X	2200 – 2500		C ≡ N
√	1700 – 1800	1700	C = O
X	1600 – 1700		C = C (aliphatic)
X	1585 – 1600		C – C (aromatic)
X	1450 – 1600		C – C (aromatic)
X	1000 – 1300		C – O
X	700 – 1000		C – X (X = Cl, Br or I)

Conclusions

The infrared spectrum appears to be aromatic and to contain N – H and carbonyl (C = O) groups.

There are, however, insufficient carbon atoms for the molecule to contain a benzene ring and other possibilities such as a substituted pyridine or pyridone ring are not possible to draw in such a way that is consistent with the molecular formula. This means that the possible aromatic C – H stretches should be treated with caution.

It is also worth noting that there are insufficient hydrogen atoms for the molecule to comprise a chain so this suggests that there is a ring containing a nitrogen atom. It cannot contain an oxygen atom as that would have to exist as an hydroxyl, O – H, group and there is no evidence for such a group.

1

Mass Spectrum

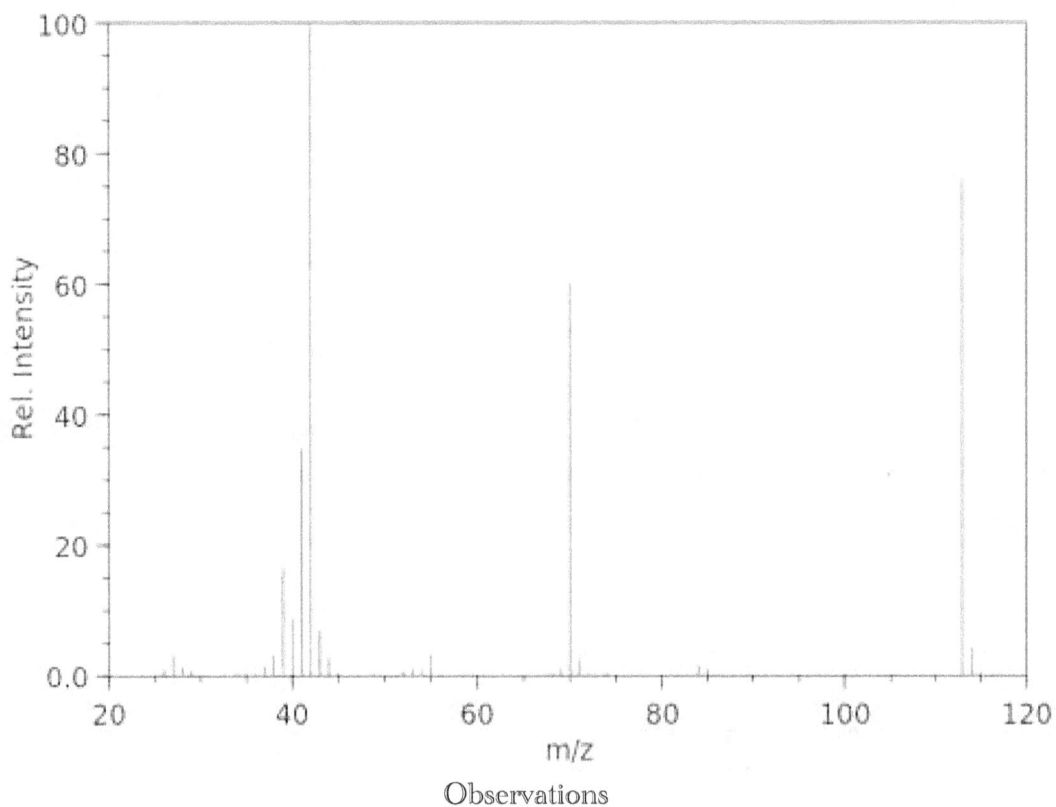

Observations

Charged fragments (m/z)	Assignment	Charged fragments (m/z)	Assignment
Molecular ion: 113	$[C_5H_7NO_2]^+$	Base peak: 42	$[C_3H_6]^+$

85	$[C_5H_6N]^+$	49	$[C_4H]_+$
70	$[C_4H_8N]^+$	41	$[C_3H_5]^+ / [C_2H_3N]^+$
56	$[C_4H_8]^+$	39	$[C_3H_3]^+$

Conclusions

We can make the following deductions:

▪ The peak at m / z = 42, assignable to $[C_3H_6]+$ suggests that there is a $- CH_2 - CH_2 - CH_2 -$ group whilst;

▪ That at m / z = 85 suggests the existence of a six membered ring containing one nitrogen atom.

Since there is no evidence for a C – O bond the two oxygen atoms, again, cannot be in the ring and since there is no evidence for a hydroxyl group this gives us the following possible structures:

I	II	III	IV	V	VI

and we can identify the correct structure from the ¹H and ¹³C nmr spectra which is our next task.

NMR Spectra

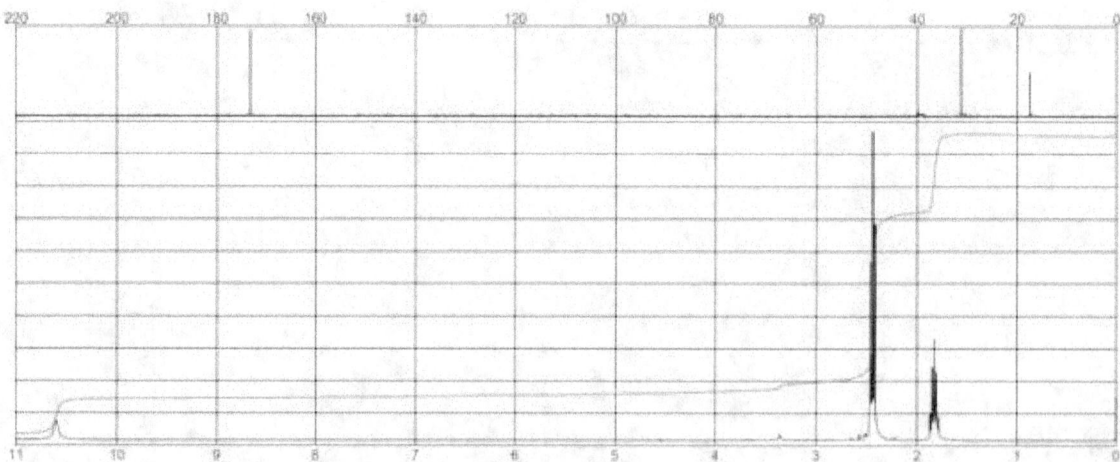

As would be expected from the molecular formula both the ¹H and ¹³C nmr contain a very limited number of peaks. We will examine the ¹³C nmr spectrum first.

¹³C NMR Spectrum

Chemical shift δ (ppm)	Integral	Assignment
175	2	C = O
32	2	C – C
17	1	C – C

It is clear from this spectrum that there is no aromatic ring since there are no signals in the δ 110 – 160 ppm region. This contradicts the infrared spectrum but this merely demonstrates that the infrared data sheet is a guide.

- There are two chemically and magnetically equivalent carbonyl groups and;
- Two chemically and magnetically equivalent C – C groups;
- One different C – C group.

This allows us to eliminate some of the original possibilities i.e. all those where the carbonyl groups would *not* be chemically and magnetically equivalent. This is delightful as that leaves us with only two possibilities:

IV   V

which both also have two chemically and magnetically C – C bonds and a distinct C – C bond.

The data sheet indicates that the carbonyl carbon atoms will occur in the region δ 160 – 220 ppm and the ¹³C nmr spectrum indicates that the two C = O carbon atoms are deshielded by appearing at δ 175 ppm. This deshielding is noticeable, and may be due to the adjaceny to the nitrogen atom but may not be significant as a shift of 15 ppm in a range of 60 ppm is not enormous.

We can, however, establish the correct structure by predicting the ¹H nmr spectrum and then examining the measured spectrum. This is our next task.

¹H NMR Spectrum

Isomer IV

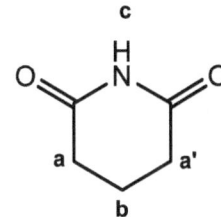

This molecule is symmetrical and H_a and $H_{a'}$ are chemically and magnetically equivalent but it is essential to consider the stereochemistry of the molecule. Due to its Π – bonding, caused by the need for the p – orbitals to overlap, the C = O bonds will be rigid, affecting the H_a and $H_{a'}$ atoms but the H_b atoms will be free to flex.

- According to the n+1 rule, H_a will be split into a triplet by H_b and this triplet will be split into a quintet by $H_{a'}$. Similarly, $H_{a'}$ will be split into a triplet by H_b and this triplet will be split into a quintet by H_a. This means that we will expect a quintet of integral *four* somewhere in the δ 0.5 – 2 ppm region..

- H_b will be split into a triplet, integral two, due to being adjacent to H_a and $H_{a'}$. There will be less splitting that for the other two pairs of hydrogen atoms as the H_b atoms will be unable to distinguish between H_a and $H_{a'}$.

- There may or may not be a signal due to the amino, N – H, hydrogen atom (H_c) of integral one. These hydrogen atoms can be labile and are observed in the spectra of some molecules and not in others.

Isomer V

For clarity, in the displayed structure, the hydrogen atoms are relabelled so that there is no confusion.

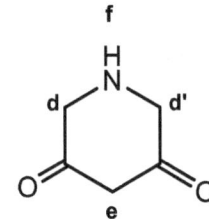

- This molecule is also symmetrical and H_d and $H_{d'}$ are mutually chemically and magnetically equivalent. There are, however, no hydrogen atoms on the adjacent carbon atoms and so these four hydrogen atoms will produce a singlet at the same resonant chemical shift.

- H_e will be also be a singlet, again due to the absence of any hydrogen atoms on adjacent carbon atoms and will be of integral two.

- Again there may or may not be a singlet, of integral one, due to the the N – H hydrogen atom (H_f).

There will, therefore, be four hydrogen atoms, $H_d/H_{d'}$, contributing to one singlet and two hydrogen atoms, H_e, which can be assigned to another singlet.

To summarise we can predict:

Chemical shift δ (ppm)	Integral	Multiplicity	Assignment
Isomer IV			
10 – 12	1	Singlet	H_c
2 – 3	4	Quintet	H_a / $H_{a'}$
0.5 – 2	2	Triplet	H_b
Isomer V			
10 – 12	1	Singlet	H_f
2 – 3	4	Singlet	H_d / $H_{d'}$
0.5 – 2	2	Singlet	H_e

If we examine the ¹H nmr spectrum, we observe a singlet of integral one at δ 11.65 ppm and if we consider the expanded spectrum in the region δ 0 – 4 ppm below:

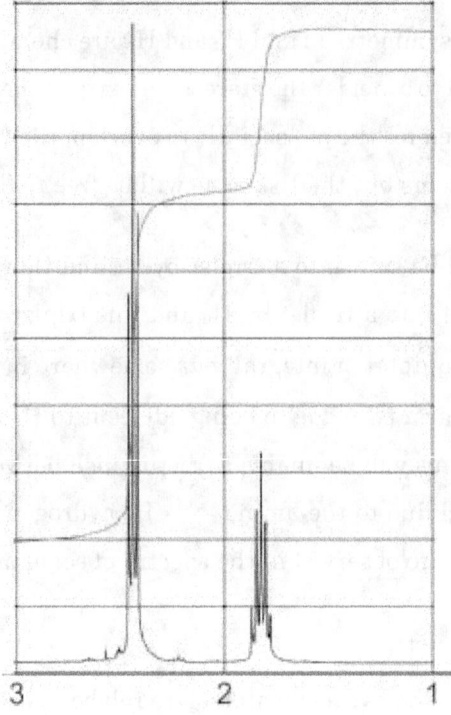

we immediately observe a triplet at δ 2.4 of integral two and a quintet of integral four at δ 1.8 ppm and so the molecule must have the structure of **isomer IV**.

Conclusions

Structure:

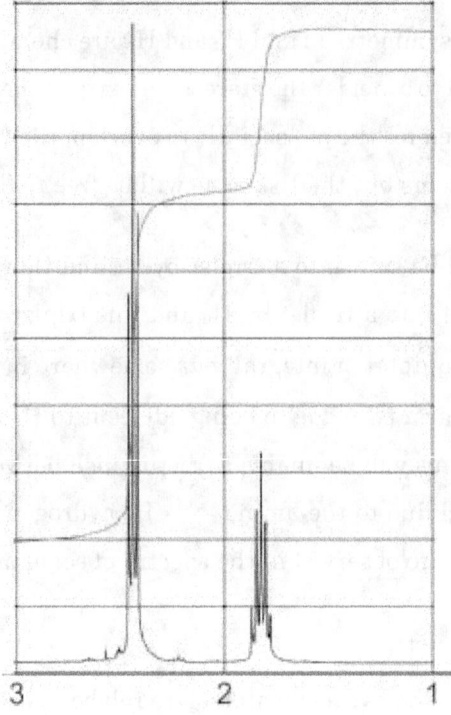

Systematic name: piperidine-2,6-dione

Chapter II

Barbituric acid

Barbituric acid is an odourless, crystalline solid which is highly soluble in water.

Although it is not, itself, pharmacologically active it is the parent compound of several thousand barbiturates which are notorious for their addictive properties but also have a role in medicine as treatments for epilepsy, anxiety and for insomnia.

It was first synthesised in 1864 by Adolf von Baeyer (Nobel Laureate in Chemistry, 1905) who also discovered the synthetic route to indigo and the precursor of the Bakelite™ polymer and also discovered the acid-base indicator, phenolphthalein which is well known to anybody who has studied chemistry. von Baeyer also developed a system of nomenclature for cyclic compounds which became incorporated into the IUPAC naming system for ring compounds.

The origin of the name of this compound is obscure but it has been suggested that von Baeyer named it after an woman named Barbara to whom he had expressed unrequited love.

With a melting point of 245°C and a boiling point of 260°C, barbituric acid is a crystalline solid. It is interesting that the difference between the melting and boiling points is so small and this suggests that the molecules display very little in the way of intermolecular interactions which will help us in the determination of its structure. It is also notable that the compound is highly soluble in water, 142g dm^{-3} at room temperature.

Barbituric acid has a formula mass of 128.09 g mol^{-1} and the following elemental composition:
C: 37.47%, H: 3.15%, N: 21.87%,O: 37.47%

This means that the **empirical** and **molecular** formulas of barbituric acid are both $C_4H_4N_2O_3$.

Infrared Spectrum

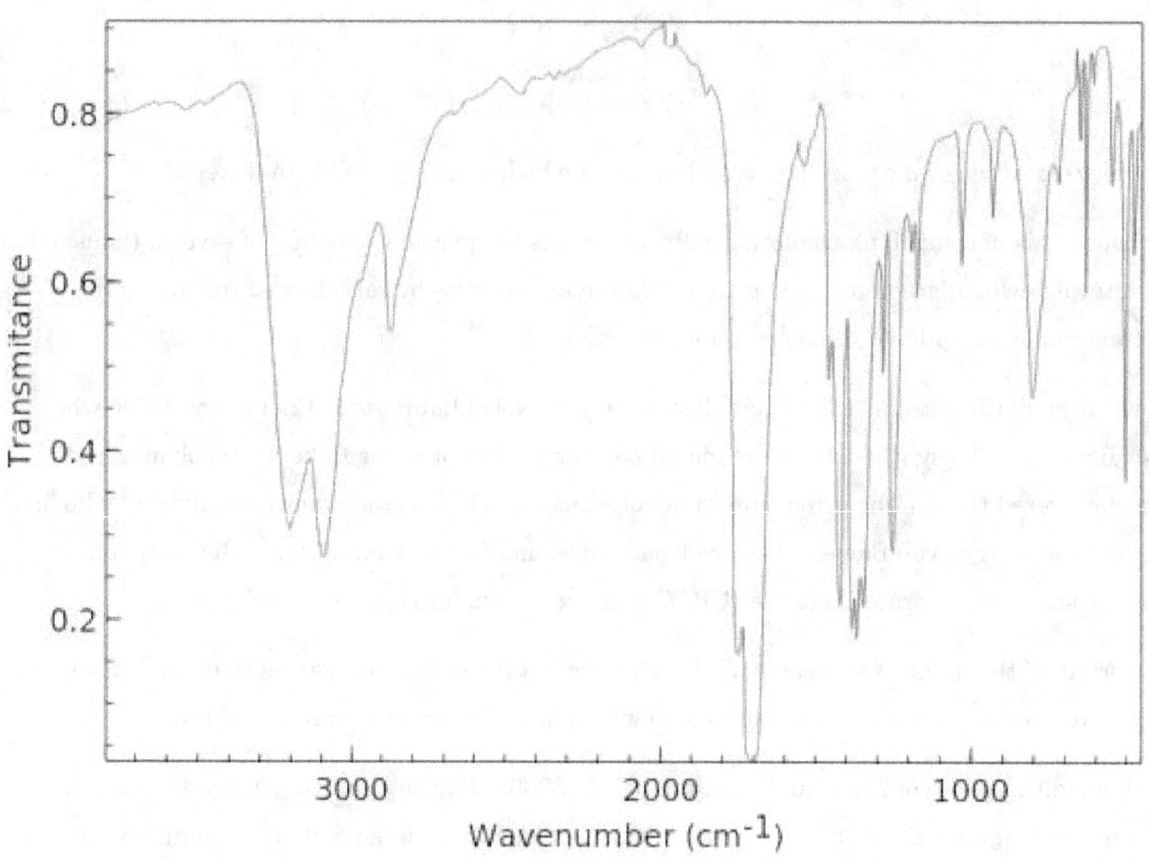

Observations

(√ / X)	Wavenumber range (cm⁻¹)	Wavenumber (cm⁻¹)	Assignment
X	3200 - 3700		O – H
√	3200 - 3600	3220	N – H
?	3000 – 3300	3090	C – H (aromatic)
√	2500 – 3000	2860	C – H (aliphatic)
X	2200 – 2500		C ≡ N
√	1700 – 1800	1720	C = O
X	1600 – 1700		C = C (aliphatic)
X	1585 – 1600		C – C (aromatic)
X	1450 – 1600		C – C (aromatic)
X	1000 – 1300		C – O
X	700 – 1000		C – X (X = Cl, Br or I)

Conclusions

▣ This compound contains at least one N – H bond and may be aromatic but the shape of the peaks above, to the left of, 3000 cm⁻¹ is not characteristic of aromatic compounds;

▣ It also contains at least one carbonyl, C = O, functional group but the absence of a C – O stretch demonstrates that the molecule is not a carboxylic acid or an ester.

Mass Spectrum

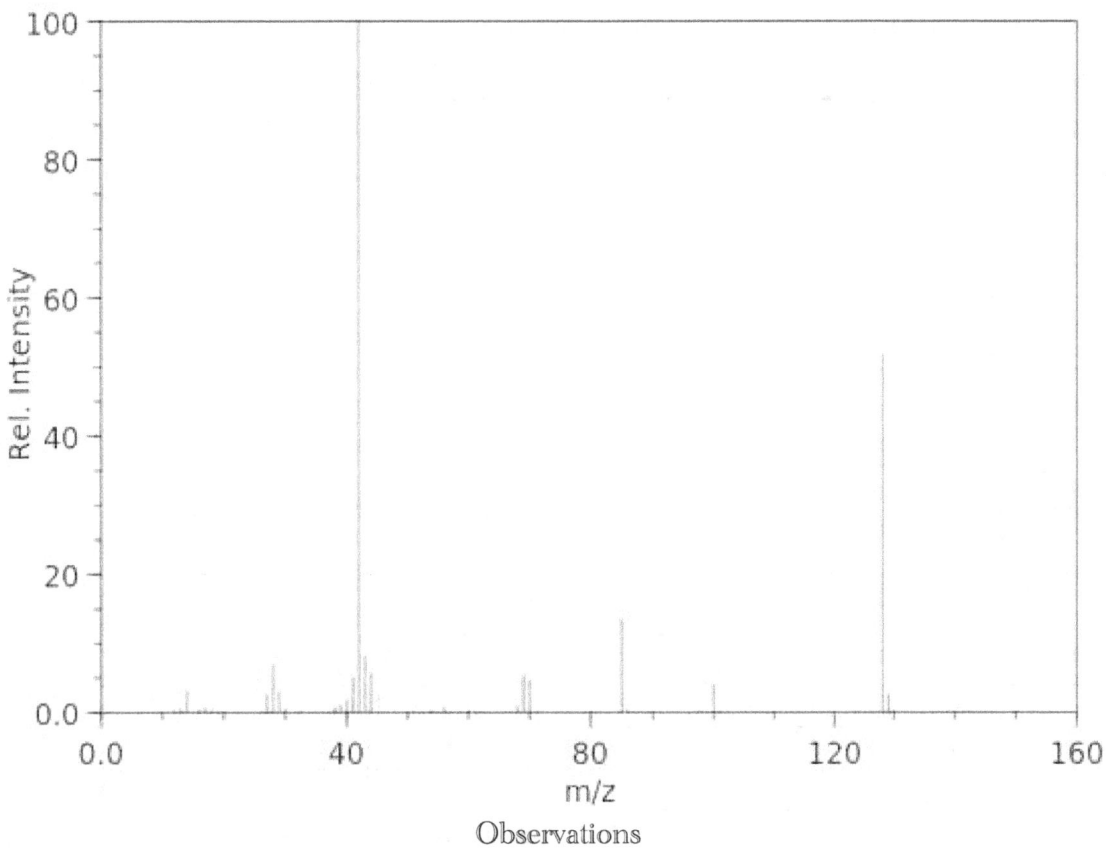

Observations

Charged fragments (m/z)	Assignment	Charged fragments (m/z)	Assignment
Molecular ion: 128	$[C_4H_4N_2O_3]^+$	Base peak: 42	$[C_2H_2O]^+$

Charged fragments (m/z)	Assignment	Charged fragments (m/z)	Assignment
100	$[C_4H_4O_3]^+$	43	$[C_2H_3O]^+$
85	$[C_3H_2O_3]^+$	39	$[C_2HN]^+$
69	$[C_2H_2O_2]^+$	28	$[CO]^+$

Conclusions

We can assume that the molecule is some sort of ring as there are insufficient hydrogen atoms for it to be linear. Since we know, from the infrared spectrum, that the molecule is not a carboxylic acid or an ester then the there must be three carbon atoms in the ring with oxygen substituents i.e. three carbonyl groups. The other three members of the ring must be the two nitrogen atoms and the remaining carbon atom existing as amino, $- NH_2$, groups and one $C - CH_2$ group. This gives us four possibilities:

I   II   III   IV

We can establish the exact structure from the ^1H and ^{13}C nmr spectra and use the infrared and mass spectra to support the determined structure.

NMR Spectra

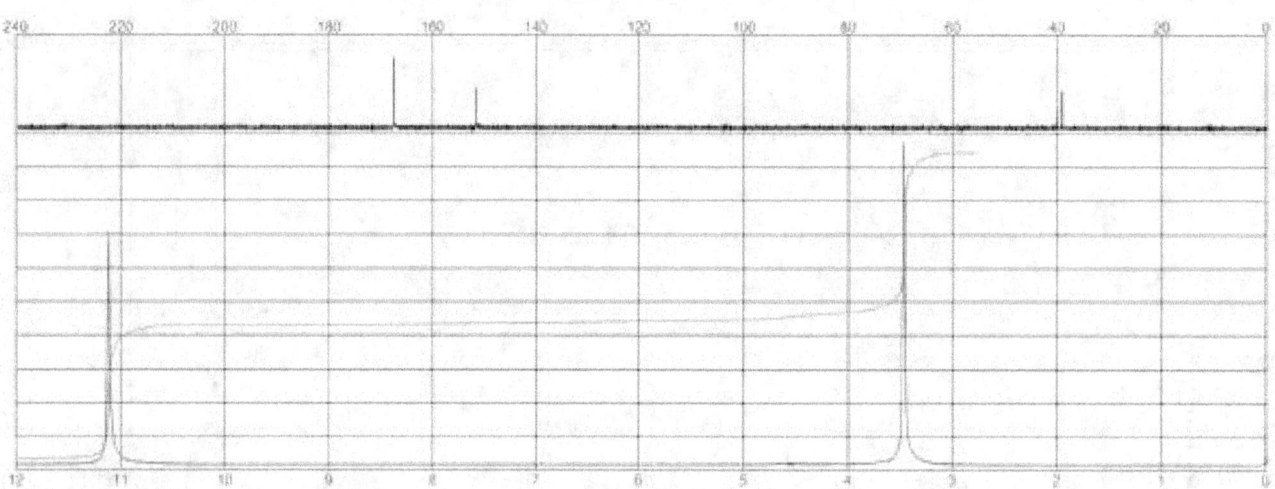

The spectra are both very simple and can be summarised, with assignments, as follows:

¹³C NMR Spectrum			
Chemical shift δ (ppm)	Integral	Multiplicity	Assignment
167	2	Singlet	C = O
153	1	Singlet	C = O
39	1	Singlet	C – H
¹H NMR Spectrum			
11.1	1	Singlet	N – H
3.45	1	Singlet	C – H

- Barbituric acid has the molecular formula $C_4H_4N_2O_3$ but it is important to remember that the integral ratios are the lowest possible whole numbers and so the two peaks in the ¹H nmr spectrum actually two chemically and magnetically amino hydrogen, N – H, atoms and two alkyl, C – H, hydrogen atoms.
- In principle, the alkyl hydrogen signal is a singlet because there are no hydrogen atoms on any adjacent carbon atoms. This rule does not apply here since amino hydrogen atoms do not couple and the other three carbon atoms are all carbonyl groups.

This confirms that the only possible structures are as shown before.

We can establish the correct structure by predicting the spectra for each candidate and comparing them to the actual observed structure.

In summary, the ¹³C **nmr spectrum** indicates that there are three types of chemically and magnetically *non-equivalent* carbon atoms in the ratio 2:1:1.

- There are two chemically and magnetically equivalent carbonyl, C=O, carbon groups and one alkyl carbon atom;
- Two of the carbonyl carbon atoms are chemically and magnetically equivalent, indicated by the integral of two, which implies there is some symmetry in the molecule;
- There is one signal of integral one which is an alkyl carbon atom.

The **¹H nmr spectrum** contains two singlets in the ratio 1:1.

These actually represent a total of four hydrogen atoms, two due to N – H atoms and two due to C – H atoms and they are both singlets which demonstrates that:

- There cannot be a sole –NH₂ group in a ring due to nitrogen's valency of three and;
- There must be a –CH₂ group. This is, of course, perfectly consistent with the molecule being cyclic but since the signal is a singlet there cannot be any hydrogen atoms on adjacent carbon atoms.

Predicting the ¹H and ¹³C NMR spectra for each isomer

Isomer I

This molecule would produce:-

- Three carbon signals,
    - Two due to carbonyl carbon atoms in the ratio in the ratio 2:1 since the molecule is symmetrical;
    - One alkyl carbon signal due to the – NH – CH₂ – NH – group at the top of the molecule.
- Two hydrogen singlets in the ratio 1 : 1 assignable to the –NH and –CH₂ groups.

This is consistent with the spectra.

Isomer II

The **¹³C nmr spectrum** will contain two carbonyl signals in the ratio 2: 1 and one alkyl signal of integral one.

The **¹H nmr spectrum** will contain three hydrogen signals in the ratio 2:1:1 since the amino, – NH groups are not chemically and magnetically equivalent.

This is inconsistent with the actual spectra so this structure can be disregarded.

Isomer III

The **¹³C nmr spectrum** will contain:-

- Two carbonyl signals in the ratio 2:1 due to the symmetry of the molecule;
- One alkyl signal of integral one.

The **¹H nmr spectrum** will contain:-

- Two singlets in the ratio 1:1 due to the symmetry of the molecule;

This is consistent with the actual, observed, spectra.

Isomer IV

In the ^{13}C **nmr spectrum** there will be three signals in the carbonyl region and one in the alkyl region.

In the ^{1}H **nmr spectrum** there will three signals in the ratio 1:1:2 since the amino, N–H, atoms are noth chemically and magnetically equivalent.

This is not what is observed and so this potential structure can also be disregarded.

This means that there are only two possible candidates for this structure,

I   III

so the question now, is, how to distinguish them.

We can achieve this by considering the electronegativity of nitrogen.

The alkyl hydrogen atoms in **Candidate I** will be significantly deshielded by their proximity to the two nitrogen atoms and will appear at the higher end of the δ 3 – 5 ppm region whilst the same atoms in **Candidate III** will not be deshielded and will appear at the lower end of the same region.

The relevant singlet in the observed ^{1}H nmr spectrum appears at δ 3.45 ppm demonstrating that there is no deshielding and so the structure must be that of Candidate III.

Conclusions

Structure:

Systematic name: pyrimidine-2,4,6(1H,3H,5H)-trione

Chapter III

Aspirin

Aspirin is a fascinating compound and the use of its precursor has been well known in many societies including ancient Sumer, pharaonic Egypt, in Roman times and across Europe in the Middle Ages across Europe.

In approximately 400BC Hippocrates referred to the efficacy of *salicylic tea* in reducing fevers whilst willow bark preparations were well known in many countries in South America.

It was, however, only in the mid eighteenth century that pharmacists, then known as apothecaries, began experimenting with the extract of willow bark for use in treating fever, specific pains such as toothache and skin inflammations.

The active ingredient in willow bark was discovered to be a compound that was later termed *salicylic acid* but it was also found to cause severe side effects such as intestinal bleeding and many patients died from the treatment due to internal bleeding.

In 1853, the chemist, Charles Frédéric Gerhardt, treated sodium salicylate with acetyl chloride (now termed ethanoyl chloride), producing acetylsalicylic acid for the first time as a synthetic compound.

It was commercialised in 1897 by the drug company, Bayer, after they had determined that acetylsalicylic acid has far fewer side effects and was marketed as *aspirin* in 1899.

Acetylsalicylic acid has a melting point of 136°C and, unusually, has a very close boiling point of 140°C, just four degrees centigrade higher upon, which, in air or oxygen, it decomposes.

The compound has the elemental composition, C: 59.95%, H: 4.48%, O: 35.51%, and a formula mass of 180.16 g mol^{-1}.

This means that the **empirical** and **molecular** formulas are both $C_9H_8O_4$.

Infrared Spectrum

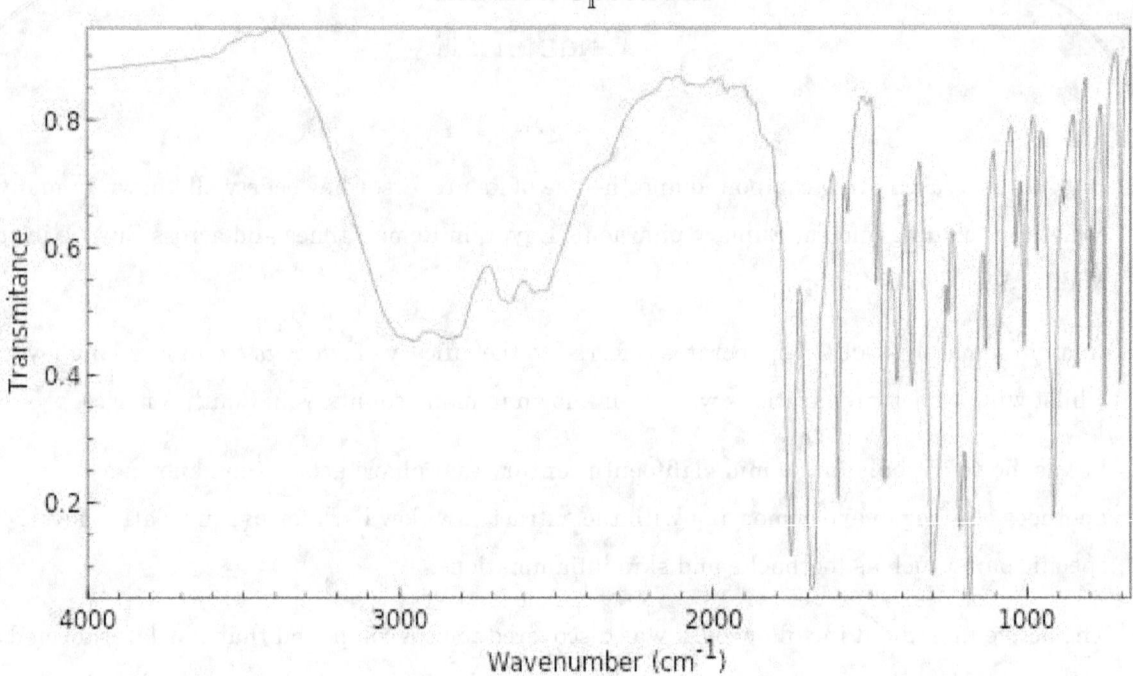

Observations

(√ / X)	Wavenumber range (cm⁻¹)	Wavenumber (cm⁻¹)	Assignment
√	3200 - 3700	3400 – 3000 (broad)	O – H
X	3200 - 3600		N – H
√	3000 – 3300	3100 – 3000	C – H (aromatic)
√	2500 – 3000	3000 – 2500	C – H (aliphatic)
X	2200 – 2500		C ≡ N
√	1700 – 1800	1780	C = O
X	1600 – 1700		C = C (aliphatic)
X	1585 – 1600		C – C (aromatic)
√	1450 – 1600	1580	C – C (aromatic)
√	1000 – 1300	1205, 1120,1080	C – O
X	700 – 1000		C – X (X = Cl, Br or I)

Conclusions

This compound is fascinating as it is both aromatic and aliphatic and contains the features of both a carboxylic acid and an ester. This is supported by the molecular formula containing four oxygen atoms which implies the existence of either two carboxylic acids, two esters or a combination of one carboxylic acid and an ester.

We can learn more from the mass spectrum (considered next) but the real detail comes from the ¹H and ¹³C nmr spectra.

Mass Spectrum

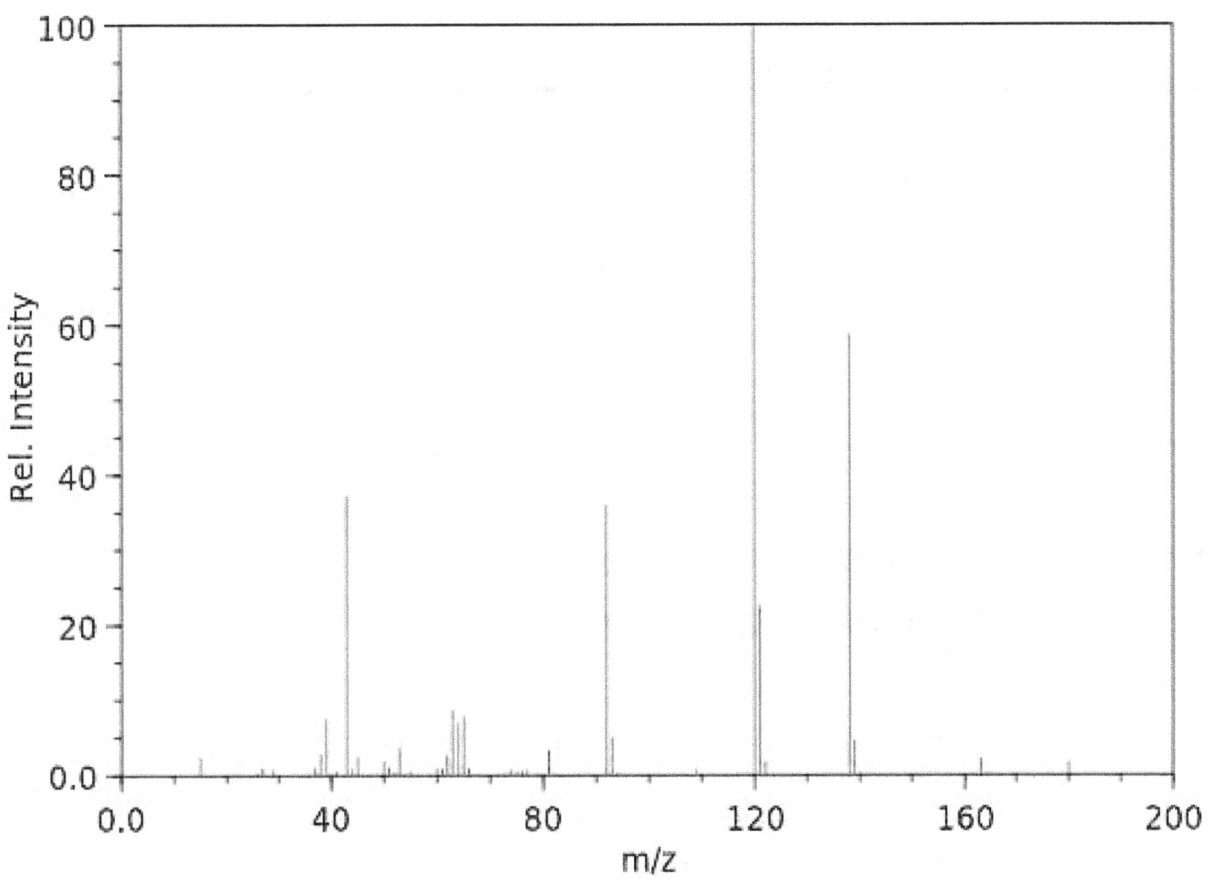

Observations

Charged fragments (m/z)	Assignment	Charged fragments (m/z)	Assignment
Molecular ion: 180	$[C_9H_8O_4]^+$	Base peak: 120	$[C_8H_8O]^+$

138	$[C_9H_6O]^+$	65	$[C_5H_5]^+$
92	$[C_7H_7]^+$	43	$[C_3H_7]^+$

Conclusions

It is difficult to draw any firm conclusions from this mass spectrum but it does appear that the molecule could be an aromatic compound with a number of substituents. Equally importantly, there is nothing to dispute the suggestions from the infra red spectrum.

We can, however, determine more from the ^{1}H and ^{13}C nmr spectra which we consider next.

NMR Spectra

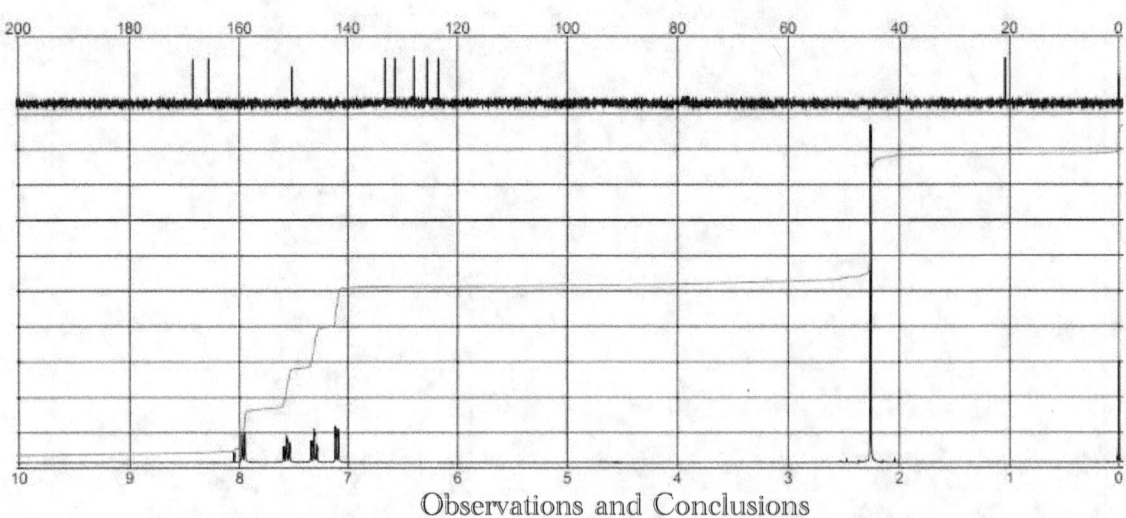

Observations and Conclusions

We will examine the ^{13}C nmr spectrum first before considering an expanded ^{1}H nmr spectrum.

Chemical shift δ (ppm)	Integral	Assignment
169	1	C = O
165	1	C = O
151	1	Aromatic C – O
135	1	Aromatic
133	1	Aromatic
129	1	Aromatic
126	1	Aromatic
123	1	Aromatic
21	1	C – C

We can deduce much information from this spectrum:

- There is one benzene ring (the aromatic peaks);
- One of the aromatic carbon atoms is bonded to an oxygen atom (δ 151 ppm);
- There is one carboxylic acid group and one ester group, accounting for all four oxygen atoms;
- The C – C bond must be part of the ester functional group and the alkyl group can only be a methyl, – CH₃, group. That alkyl group cannot be bigger as there would be more than one C – C signal.

This gives us three possible isomers:

I   II   III

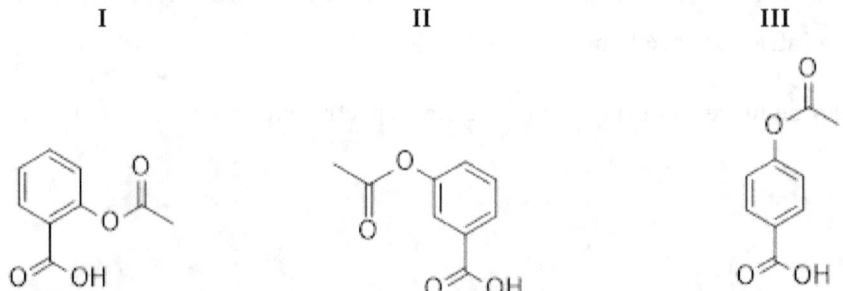

We can identify the precise isomer from the ^{1}H nmr spectrum which we consider next.

13

^{1}H NMR Spectrum

There is no need to consider the singlet in any detail as it is the alkyl region and, with an integral of three is clearly due to the – CH$_3$ group and so we only need to concentrate on the aromatic region. This alkyl group must be bonded to a carbon atom with no hydrogen atoms on the adjacent carbon atom(s).

This means that we can consider the aromatic region in detail.

At a quick glance, the region might appear to comprise two doublets and two triplets however, upon expansion, as shown below the peaks are actually two doublets of doublets and two triplets of doublets. The expanded region is shown below.

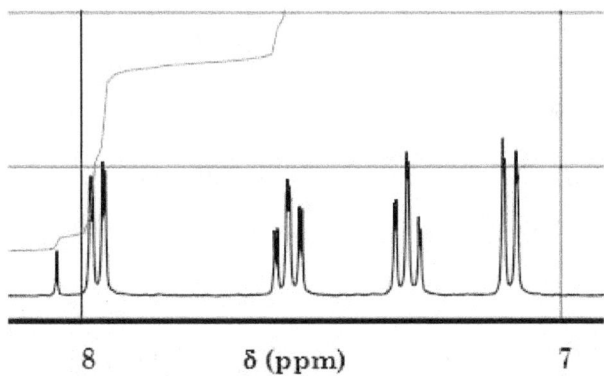

The peak at δ 8.2 ppm can be disregarded as its integral is a fraction of one and can be regarded as an impurity and the peaks are tabulated below.

Chemical shift δ (ppm)	Multiplicity	Integral	Assignment
7.85	Doublet of doublets	1	Aromatic
7.60	Triplet of doublets	1	Aromatic
7.30	Triplet of doublets	1	Aromatic
7.10	Doublet of doublets	1	Aromatic
2.30	Singlet	3	Methyl group

This is an intriguing example of the application of high resolution ^{1}H nmr spectroscopy as the multiplicities are governed by the **n+1** rule. In aliphatic molecules, **n** is the integral number of hydrogen atoms on adjacent carbon atoms whereas in aromatic compounds there can be, by their very nature, communication between atoms further apart.

The greater the number of carbon atoms, however, the weaker communication which explains why there are varying splittings in the multiplet.

The splitting (J) is the distance between peaks in the multiplet and is measured in Hertz (Hz). If we consider one of the doublets, the distance (splitting between the two larger central peaks) is greater than the splitting within the doublet itself, indicating that, although hydrogen atoms bonded to carbon atoms which are not immediately adjacent, the communication reduces with distance. The same applies to all multiplets.

If we consider one of the doublets then we can determine from the two large central peaks that there is one hydrogen on adjacent carbon atoms and then there must be another hydrogen atom on the next carbon atom as shown below.

With triplets, the three central peaks indicate that there is a total of two hydrogen atoms on adjacent atoms and, since the central peaks are split into doublets this indicates that there is one hydrogen atom bonded to the next carbon atom.

We will investigate the three candidate isomers in turn using alphabetically labelled hydrogen atoms.

Isomer I

Explaining the formation of the two triplets of doublets

Considering the signal due to H_b,

H_b has two hydrogen atoms on adjacent carbon atoms so will form a triplet (as shown below left) and this triplet will be split into a doublet of triplets by the more distant H_d as shown:

The same applies to H_c which is split into a triplet by H_d and H_c with the triplet then split into a triplet of doublets by H_d. Since there is only one hydrogen on each of the four carbon atoms, this accounts for the two triplets of doublets of integral one.

Explaining the formation of the two doublets of doublets

Considering Ha (H_d will behave similarly) we can explain the doublet of doublets as shown:

H_a is split into a doublet by H_b and this doublet is split into a doublet of doublets by H_c.

This doublet will be split into a doublet of doublets of doublets by H_d but the splitting is so small as to not be measurable – remember that the splitting decreases rapidly with increasing distance.

This accounts for the ^1H nmr spectrum and supports the conclusions from the ^{13}C nmr spectrum and this structure is clearly correct.

It is, however, an interesting exercise to consider the other possibilities which we will now do briefly.

Isomer II

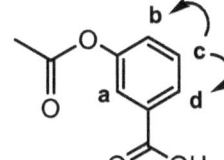

Excluding the hydrogen atom symbols, we can see that

■ H_a would cause a singlet since there are no hydrogen atoms on adjacent carbon atoms;

■ H_b would give rise to a doublet of doublets due to splitting by H_c and then by H_d. Likewise, H_d will produce a doublet of doublets due to splitting by H_c and then by H_b.

H_b will be split into a doublet by H_c which is split into a doublet of doublets by H_d.

■ H_c will produce a triplet due to simultaneous splitting by H_b and H_d.

H_c will be split into a triplet due to splitting by H_b and by H_d.

In summary, then, this molecule will produce one singlet, two doublets of doublets and a triplet (all of integral one) and since this is not observed this structure cannot be correct.

Isomer III

Isomer III has the following proposed structure.

This molecule is symmetrical hence the aromatic hydrogen atoms are labelled a / a' and b / b' respectively. Since all of the hydrogen atoms have only one hydrogen atom on adjacent atoms this means that the ¹H nmr spectrum will comprise two doublets of integral two.

This occurs because

■ H_a will be split by H_b and H_b will be split by H_a.

■ Likewise $H_{a'}$ will be split by $H_{b'}$ and $H_{b'}$ will be split by $H_{a'}$.

Note that there will be no splitting by hydrogen atoms on the opposite side of the molecule, for example no splitting of H_a's signal by $H_{b'}$ due to the absence of hydrogen atoms on two of the intermediate carbon atoms and so the n+1 rule does not apply.

Conclusions

Structure:

Systematic name: 2-acetoxybenzoic acid

Trivial name: Acetylsalicylic acid.

Chapter IV

Methyl paraben

Methyl paraben is the simplest of a class of compounds known as parabens which are used as preservatives in pharmaceuticals as well as in cosmetics because they have anti-bacterial and fungicidal properties. They are also used as food preservatives.

Methyl paraben is found naturally in some fruits such as blueberries but now, used in many daily daily and night-time facial products, it is usually produced synthetically. Very occasionally, it can can cause allergies but this is extremely rare.

Methyl paraben has a melting point of 125°C and a boiling point of 275°C.

The compound has the elemental composition, C: 63.10%, H: 5.31%, O: 31.55%, and a formula mass of 152.15 g mol^{-1}.

This means that the **empirical** and **molecular** formulas are both $C_8H_8O_3$.

The presence of three oxygen atoms implies the existence of either a carboxylic acid or an ester with one additional oxygen atom somewhere else in the structure.

Infrared Spectrum

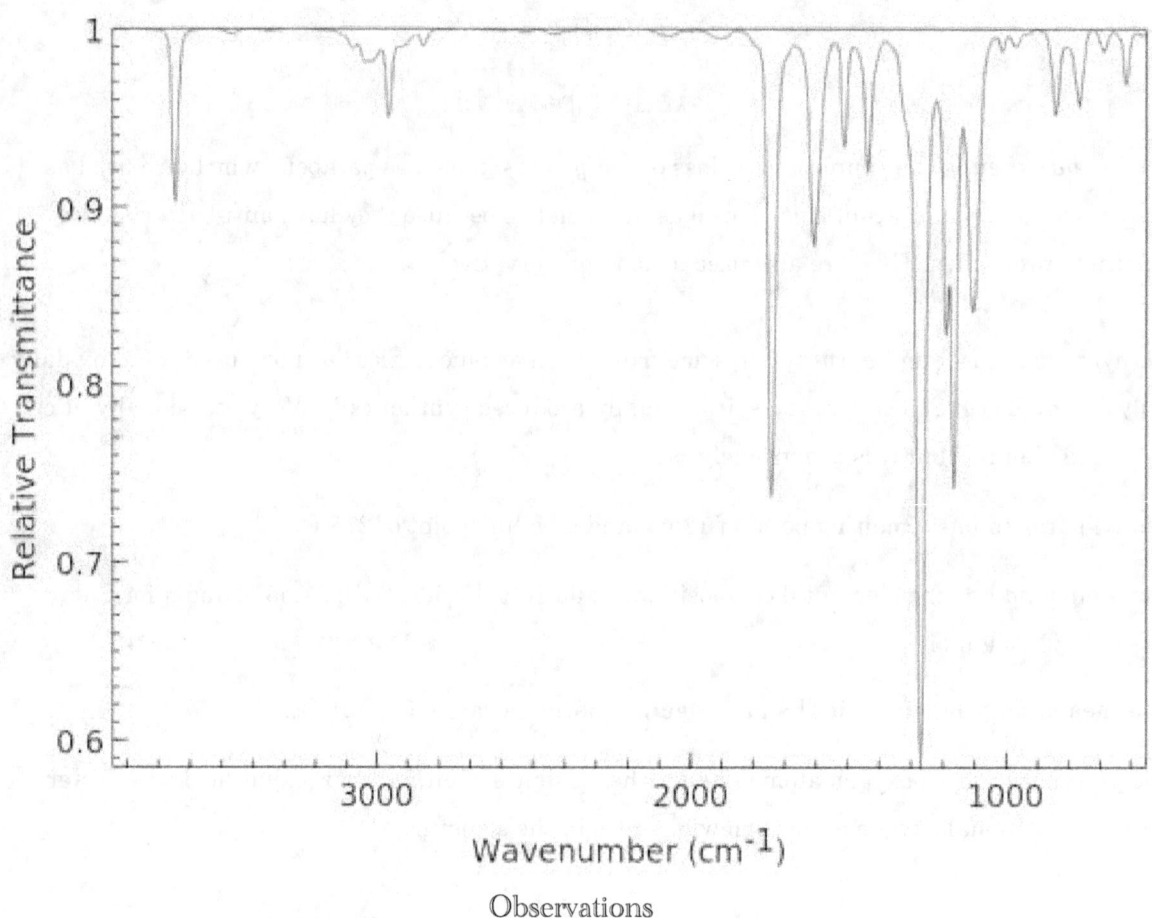

Observations

(√ / X)	Wavenumber range (cm⁻¹)	Wavenumber (cm⁻¹)	Assignment
√	3200 - 3700	3620	O – H
X	3200 - 3600		N – H
X	3000 – 3300	3160, 3100	C – H (aromatic)
√	2500 – 3000	2950	C – H (aliphatic)
X	2200 – 2500		C ≡ N
√	1700 – 1800	1740	C = O
X	1600 – 1700		C = C (aliphatic)
X	1585 – 1600		C – C (aromatic)
X	1450 – 1600		C – C (aromatic)
√	1000 – 1300	1280	C – O
X	700 – 1000		C – X (X = Cl, Br or I)

Conclusions

■ This compound is aromatic and the ring has an aliphatic group substituent.

■ The stretches at 3620 cm⁻¹ and 1740 cm⁻¹ indicate that it is both an alcohol and an ester or a carboxylic acid.

Mass Spectrum

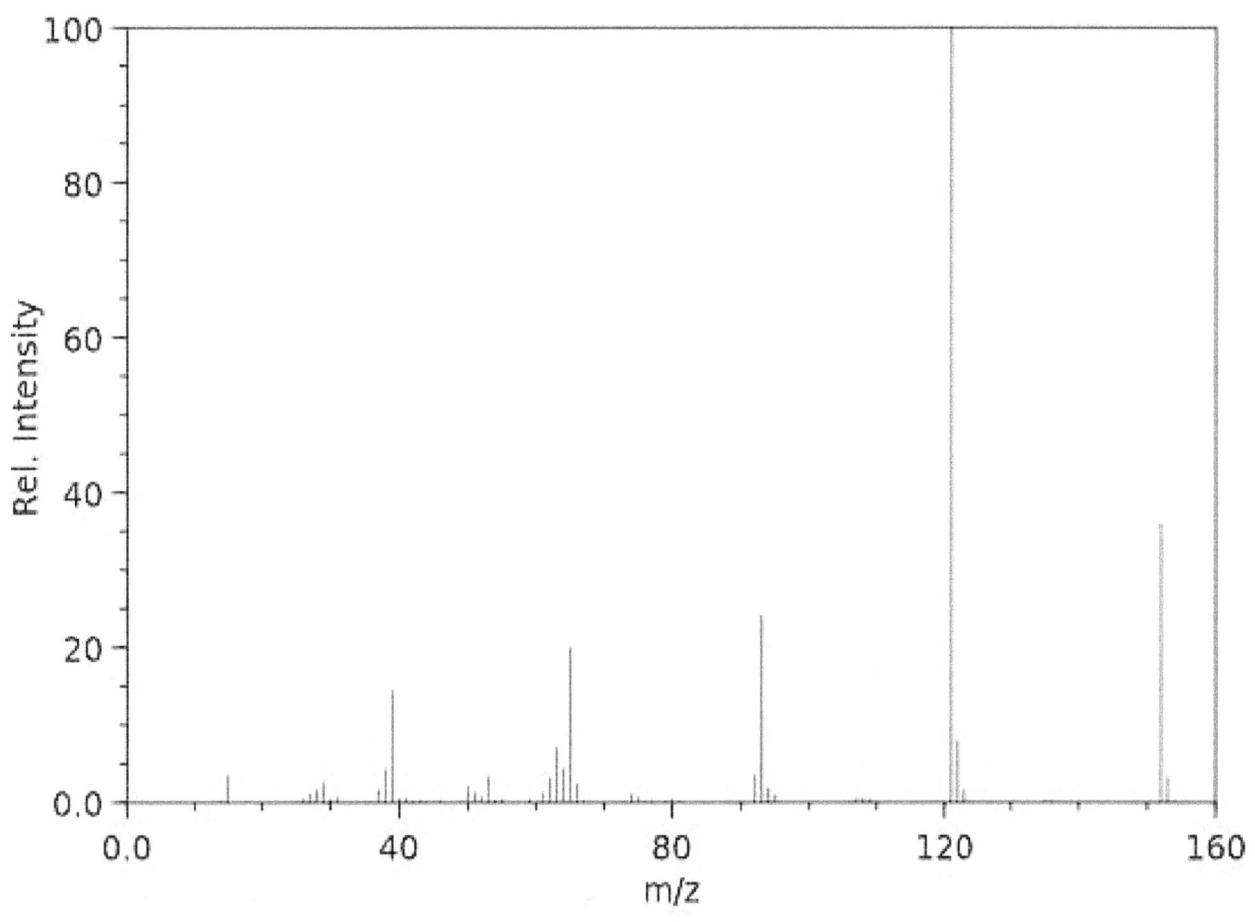

Observations

Charged fragments (m/z)	Assignment	Charged fragments (m/z)	Assignment
Molecular ion: 152	$[C_8H_8O_3]^+$	Base peak: 121	$[C_7H_5O_2]^+$

93	$[C_7H_9]^+$	39	$[C_3H_3]^+$
65	$[C_5H_5]^+$	29	$[C_2H_5]^+$
28	$[CO]^+$	16	$[OH]^+$

Conclusions

It is difficult to conclude anything much from this spectrum beyond:-

- The possibility of the presence of a six-membered carbon ring although there are none of the peaks characteristic such a ring;
- The difference between the molecular ion and the base peak is equal to m/z = 31 which can be assigned to –OCH₃, lending support to the concept that the molecule is an ester although it is possible that this group is directly substituted on a ring;
- The peak at m/z = 28 could also be assigned to part of an ester group;
- The four peaks at m/z = 29, 39, 65 and 93 could also be due to an aliphatic chain.

We can learn more from the ¹H and ¹³C nmr spectra.

NMR Spectra

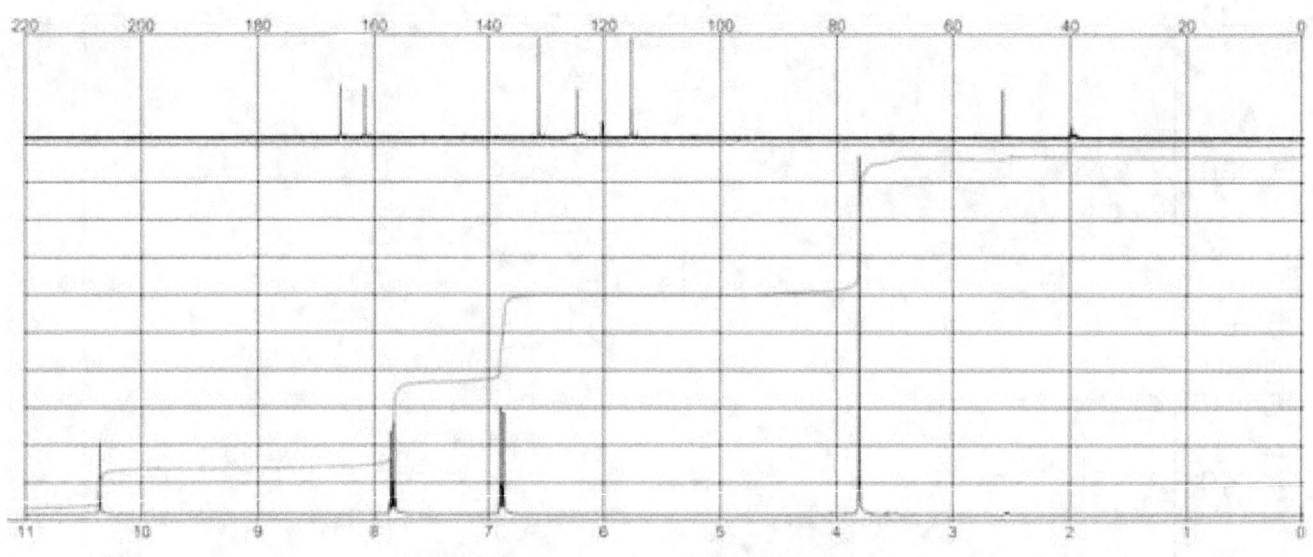

Observations and Conclusions

Both of these spectra are very clear and it makes little difference which we consider first so we choose to examine the ^{13}C nmr spectrum first as that spectrum comprises singlets whilst the ^1H nmr spectrum demonstrates multiplicities which can provide even more detail.

^{13}C NMR Spectrum

Chemical shift δ (ppm)	Integral	Assignment
166	1	Carbonyl
162	1	Aromatic carbon with an OH group
132	2	Aromatic carbon
124	1	Aromatic carbon
116	2	Aromatic carbon
52	1	Methyl

Conclusions

- It is quite clear that this compound contains an aromatic ring which must be a benzene ring.
- From the infrared and mass spectra we can surmise that this molecule has an ester functional group which, given the number of carbon atoms, implies the existence of a $- CO_2CH_3$ group bonded to the ring.
- Given the limited number of hydrogen atoms the $-$ OH functional group must also be attached to the ring.

This gives us the following possible candidate structures which can be distinguished by the ^1H nmr spectrum:

I II III

^{1}H NMR Spectrum

From the ^{13}C nmr, we have concluded that the molecule comprises a:

- Benzene ring with two functional groups:
    - An – OH group and
    - An ester grouping of formula – CO_2CH_3. There cannot be a larger ester group than the methyl, – CH_3, group on the ester functional group for two reasons:
        - Insufficient carbon atoms in the molecular formula.
        - The presence of a single peak in the alkyl region of the ^{13}C nmr spectrum.

and produced three possible candidates for the structure. This also means that, using the data sheet at the beginning of this volume, we can roughly predict the ^{1}H nmr spectrum as follows:

- Firstly, but least importantly, there may or may not be a peak due to the – OH hydrogen since hydroxyl hydrogen atoms are labile and may not be detected. If it appears then it will be of integral one.
- There will be a peak of integral three in the alkyl region and, since there are no hydrogen atoms on the adjacent carbon, it will be a singlet.
- There will be two or three peaks in the aromatic region, totalling an integral of four.

Before examining the ^{1}H nmr spectrum we will predict the *aromatic* regions only of all three isomers on the basis of a low resolution spectrum and this is our next task.

Isomer I

There is no symmetry in the molecule and so the aromatic region will contain four specific peaks each of integral one. There will however, be couplings between the aromatic hydrogen atoms and this can be summarised below using the labelling in the structure below:. Working anti clockwise,

This doublet of doublets is split into a doublet of doublets of doublets by Hd

This doublet is split into a doublet of doublets by Hc

Ha is split into a doublet by Hb

Similarly, H_b will be split into a doublet by Ha and this doublet will be split into a doublet of doublets by Hc and this will be split into a doublet of doublets of doublets by Hd.

Hc and Hd will be split in a similar manner. This pattern is not observed and so this isomer cannot be the molecule we are examining.

Isomer II

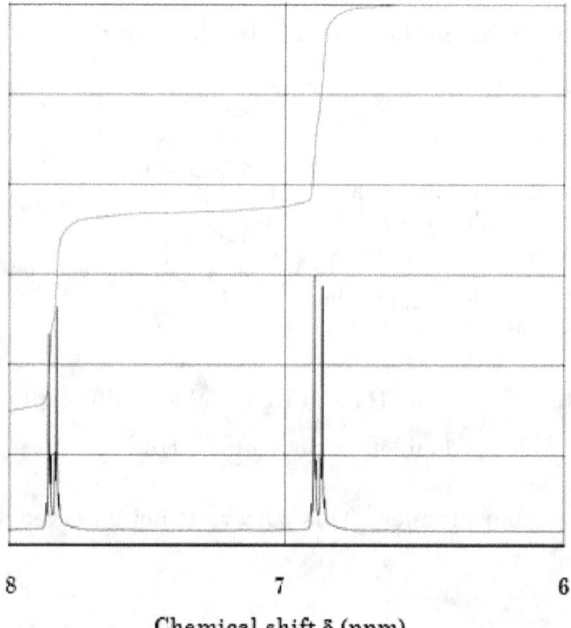

Using new labelling,

- H_a will be a singlet as there are no hydrogen atoms on immediately adjacent carbon atoms;
- H_b will be split into a doublet by H_c;
- This doublet will be split into a doublet of doublets by H_d.

H_c and H_d will be split similarly and, again, since this is not observed, Isomer II cannot be the correct structure.

Isomer III

This molecule is symmetrical hence the numbering a / a' and b / b'.

H_a will be split into a doublet by H_b whilst $H_{a'}$ will be split into a doublet by $H_{b'}$.

This means that we would observe two doublets. Unlike the previous candidates would display four signals this molecule, will have integrals of two for each of the two doublets.

This is exactly what the ^{1}H nmr spectrum (expanded region shown below) displays.

8   7   6

Chemical shift δ (ppm)

Conclusions

From the ^{1}H and ^{13}C NMR spectrum we can make the following assignments.

^{13}C nmr spectrum:

Chemical shift δ (ppm)	Integral	Assignment
166	1	C_c
162	1	C_f
132	2	C_a and $C_{a'}$
124	1	C_b and $C_{b'}$
116	2	C_e
52	1	C_d

^{1}H nmr spectrum:

Chemical shift δ (ppm)	Integral	Multiplicity	Assignment
3.80	3	Singlet	H_g
6.90	2	Doublet	H_h / $H_{h'}$
7.85	2	Doublet	H_i / $H_{i'}$
11.35	1	Singlet	H_j

Questions could be asked about the assignments of H_h/$H_{h'}$ and H_i/$H_{i'}$ but since H_i/$H_{i'}$ are closest to the electronegative oxygen atom they are deshielded and will appear at higher chemical shifts.

Overall Conclusions

Structure:

Systematic name: methyl 4-hydroxybenzoate

Chapter V

Ethyl salicylate

Ethyl salicylate is used in the treatment of sports related pain and muscular strains. It is also used in perfumery and artificial flavourings.

With a melting point of 1.3°C and a boiling point of 232°C it is a colourless liquid at room temperature. Sparingly soluble in water, it dissolves readily in ethanol and diethyl ether.

This compound has a very pleasant, mint-like, odour resembling the odours extracted from *wintergreen* plants and is used in perfumes, mouthwash, toothpaste and artificial flavours.

Wintergreen is a term for a group of aromatic plants but it has been largely replaced with the term *evergreen*. Wintergreen berries have been used for centuries medically and Native Americans use it to brew an infusion from the leaves to alleviate rheumatic symptoms, headache, fever, sore throat, and various aches and pains.

It has been suggested that these, proven, therapeutic effects arise because the primary metabolite is salicylic acid, the precursor of aspirin (Chapter III).

Ethyl salicylate has the elemental composition, C: 64.99%, H: 6.08%, O: 28.89%, and a formula mass of 166.17 g mol^{-1}.

This means that the **empirical** and **molecular** formulae are both $C_9H_{10}O_3$.

The presence of three oxygen atoms implies the existence of either a carboxylic acid or an ester with one additional oxygen atom somewhere else in the structure.

V - Ethyl salicylate

Infrared Spectrum

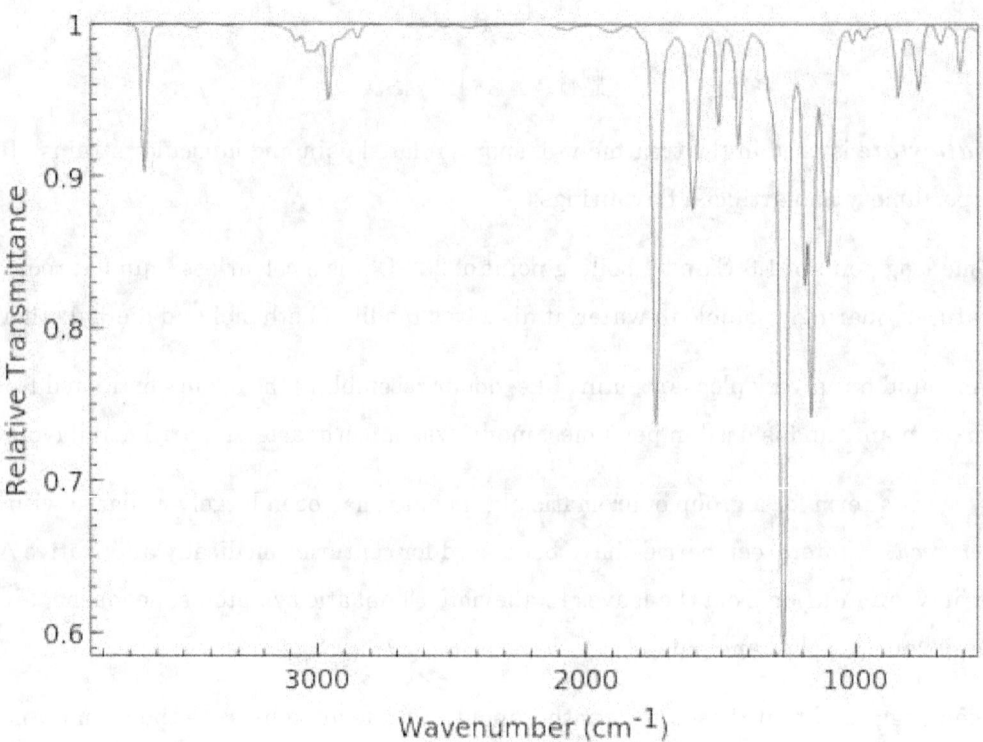

Observations

(√ / X)	Wavenumber range (cm⁻¹)	Wavenumber (cm⁻¹)	Assignment
√	3200 - 3700	3620	O – H
X	3200 - 3600		N – H
√	3000 – 3300	3060	C – H (aromatic)
√	2500 – 3000	2970	C – H (aliphatic)
X	2200 – 2500		C ≡ N
√	1700 – 1800	1755	C = O
X	1600 – 1700		C = C (aliphatic)
X	1585 – 1600		C – C (aromatic)
X	1450 – 1600		C – C (aromatic)
√	1000 – 1300	1290	C – O
X	700 – 1000		C – X (X = Cl, Br or I)

Conclusions

- This compound is aromatic and, given the molecular formula, it is reasonable to assume that the aromatic ring is six-membered.

- Since the molecule contains an hydroxyl, – OH group; a carbonyl, – C = O group and a C – O bond it is reasonable to propose that it contains a carboxylic acid functional group or an ester functional group with a separately bonded – OH group bonded to the ring.

Mass Spectrum

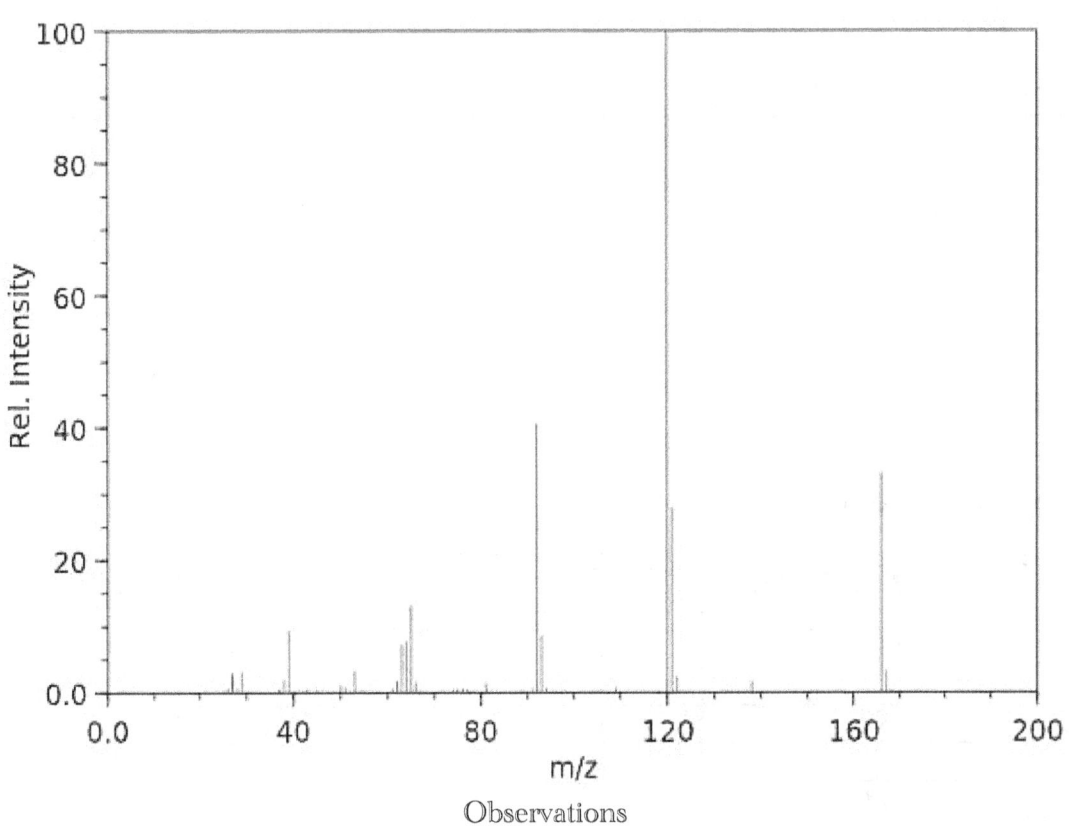

Observations

Charged fragments (m/z)	Assignment	Charged fragments (m/z)	Assignment
Molecular ion: 166	$[C_9H_{10}O_3]^+$	Base peak: 120	$[C_8H_8O]^+$
92	$[C_7H_8]^+$	39	$[C_3H_3]^+$
65	$[C_5H_5]^+$	29	$[C_2H_5]^+$

Conclusions

This spectrum does not really help too much other than to indicate:-

▪ That the molecule is likely to contain a benzene ring and the remaining atoms forming substituents whilst

▪ The tiny peak at m/z = 45 can be assigned to either $[OCH_2CH_3]^+$ or $[COOH]^+$ which does not help distinguish between the presence of a carboxylic acid or ester functional group.

We can establish the exact structure from the ^1H and ^{13}C nmr spectra which we consider next.

NMR Spectra

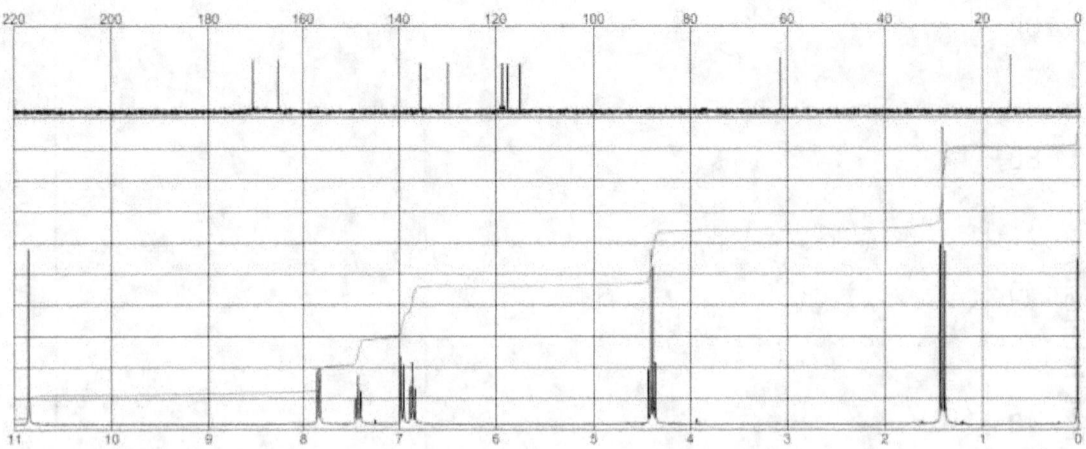

Our first task is to analyse the ^{13}C nmr spectrum:

Chemical shift δ (ppm)	Integral	Assignment
171	1	C = O
165	1	Aromatic carbon C – OH
135	1	Aromatic C
130	1	Aromatic C
119	1	Aromatic C
118	1	Aromatic C
116	1	Aromatic C
62	1	C – O
14	1	C – C

Observations and Conclusions

▦ There is clearly a benzene ring with an – OH substituent;

▦ The remaining carbon, hydrogen and oxygen atoms must constitute an ester grouping which we know, from the infra red spectrum, could exist. If we assume that this group forms one substituent then we have a benzene ring with one – OH and an ester group, R, which means that we have three possible structures:

If we total up the atoms already present we find that the base formula is $C_7H_5O_3$ which leaves us with C_2H_5 to fit in. This immediately identifies – R as – CH_2CH_3 and this accords with the peaks at ⬚ 171, 62, 14 and this means that we can now draw the three possible isomers as:

I

II

III

We can determine the correct structure of the isomer using the ^{1}H nmr spectrum.

^{1}H NMR Spectrum

- The peak at δ 10.9 ppm (integral one) is clearly due to the – OH group;
- The quartet at δ 4.4 ppm (integral two) must be due to the – OCH_2 – CH_3 for two reasons:
  - It is in the O – CHx region
  - It must be adjacent to carbon atoms with a total of three hydrogen atoms (n+1 rule). In this group there is only one other carbon and so this must be the – OCH_2 – CH_3 hydrogen atoms.
- The triplet at δ 1.4 ppm (integral 3) must be due to a methyl, – CH_3, group bonded to a – CH_2 group.

This leaves us to determine the isomer by considering the arrangements of the *aromatic* hydrogen atoms.

Despite initially appearing complex, this spectrum is straightforward to analyse.

- **Isomer III** can be immediately discounted since, being symmetrical, this region would comprise two pairs of doublets each of integral two. which do not appear in the spectrum.

- **Isomer II**

Since there will be a singlet caused by the circled hydrogen atom we need go no further with this possibility and, if our conclusions so far are correct then this leaves us with Isomer I as the only possibility.

- **Isomer I**

It is not sufficient to decide that the structure is the candidate remaining and we need to explain the spectrum in detail. This isomer is shown again with the relevant hydrogen atoms labelled alphabetically:

We will consider the multiplicity of each signal in turn:

- H_a will be split into a **doublet of doublets of doublets** by sequential splitting by H_b, H_c and then H_d:

Hb splits the Ha signal into a doublet

This doublet is split into a doublet of doublets by Hc

Hd splits the doublet of doublets into a doublet of doublets of doublets

- H_b will be split into a triplet by H_a and H_b and this triplet is then split into a **doublet of triplets** by H_d:

Hb is split into a triplet by Ha and Hc

This triplet is split into a doublet of triplets by Hd

- In the same way, H_c will be split into a triplet by H_b and H_d and this triplet is then split into a **doublet of triplets** by H_a:

This triplet is split into a doublet of triplets by Ha

Hc is split into a triplet Hc and Hd

- Finally, we must consider the signal caused by H_d.
  - As with H_a, H_d is split into a doublet by H_c;
  - This doublet is split into a doublet of doublets by H_b which is
  - Split into a doublet of doublets of doublets by H_a:

Ha splits the doublet of doublets into a doublet of doublets of doublets

This doublet is split into a doublet of doublets by Hc

Hd is split into a doublet by Hc

This fully accounts for the aromatic portion of the ^1H spectrum.

Conclusions

Structure:

Systematic name: ethyl 2-hydroxybenzoate

Chapter VI

Mandelic acid

First discovered in 1831 by Ferdinand Ludwig Winckler (1801–1868) while heating an extract of bitter almonds, amygdalin, with dilute hydrochloric acid, *mandelic acid* is used as an anti-bacterial treatment and is of especial use in the treatment of urinary tract infections. It has also been used as an oral antibiotic and is used in chemical face peels.

Its name is derived from *mandel* which is the German word for almond.

Mandelic acid is optically active and the angle of rotation is +/- 158°. It is prescribed as a racemic mixture but is also produced naturally via enzymes such as mandelate racemase which interconverts the two enantiomers. Derivatives of mandelic acid form through the metabolism of adrenaline and noradrenaline and it is also a by-product of the biodegradation of styrene. Phenylpyruvic acid which is derived from phenylalanine, a naturally occurring essential α-amino acid is another precursor to mandelic acid.

The racemic mixture of mandelic acid has a melting point of 119°C and a boiling point of 322°C which means that it exists as a colourless crystalline solid. The individual, purified, enantiomers have a melting point of 132°C.

Mandelic acid has the elemental composition of: C: 63.10%, H: 8.08%, O: 31.55% and a formula mass (M_r) of 152.15 g mol^{-1} and so this means that the **empirical** and **molecular** formulas are both $C_8H_8O_3$.

Infrared Spectrum

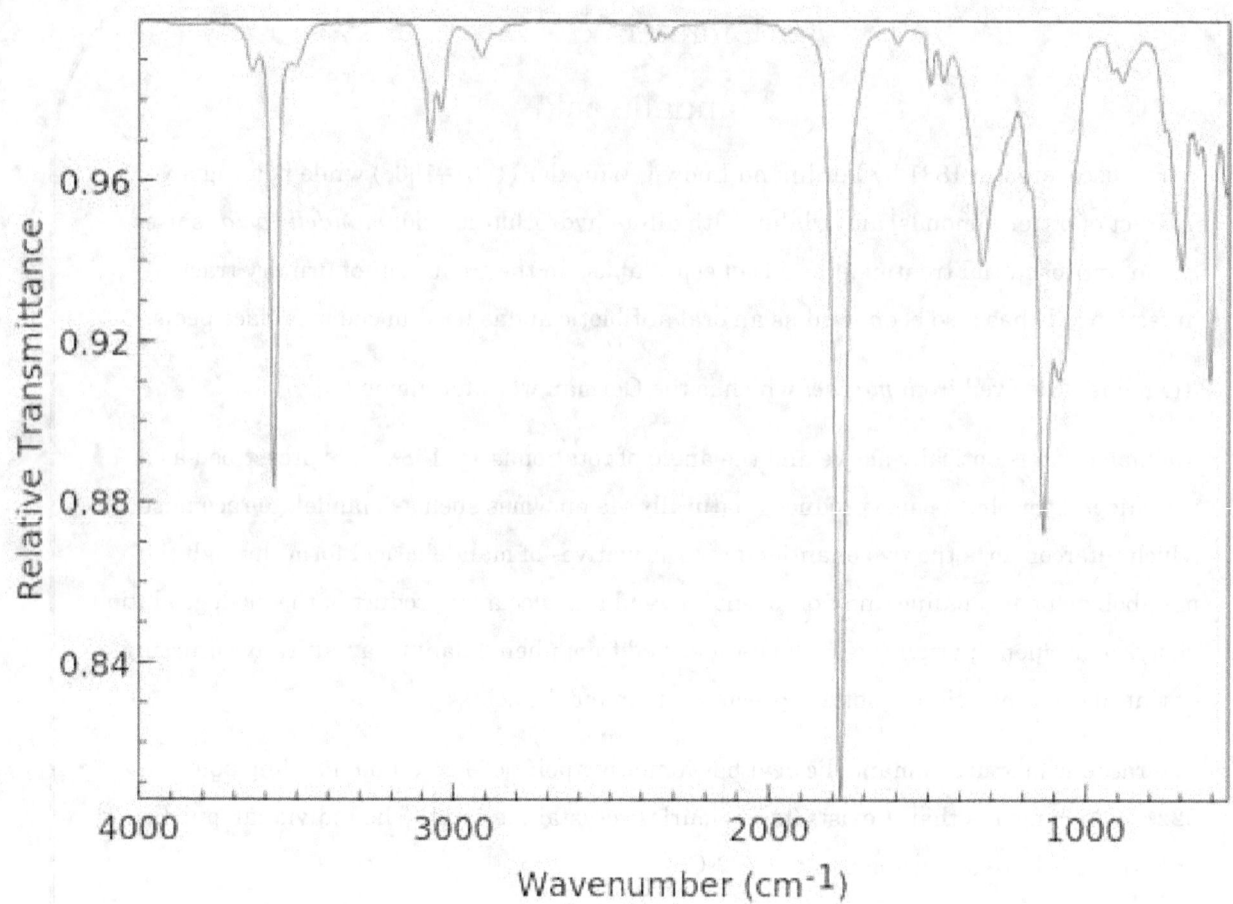

Observations

(√ / X)	Wavenumber range (cm⁻¹)	Wavenumber (cm⁻¹)	Assignment
√	3200 - 3700	3580	O – H
X	3200 - 3600		N – H
√	3000 – 3300	3080, 3010	C – H (aromatic)
√	2500 – 3000	2900	C – H (aliphatic)
X	2200 – 2500		C ≡ N
√	1700 – 1800	1780	C = O
X	1600 – 1700		C = C (aliphatic)
X	1585 – 1600		C – C (aromatic)
X	1450 – 1600		C – C (aromatic)
√	1000 – 1300	1120	C – O
X	700 – 1000		C – X (X = Cl, Br or I)

Conclusions

This compound:-

- Is an aromatic alcohol with an aliphatic grouping and / or;
- Also contains a carboxylic acid functional group.

Mass Spectrum

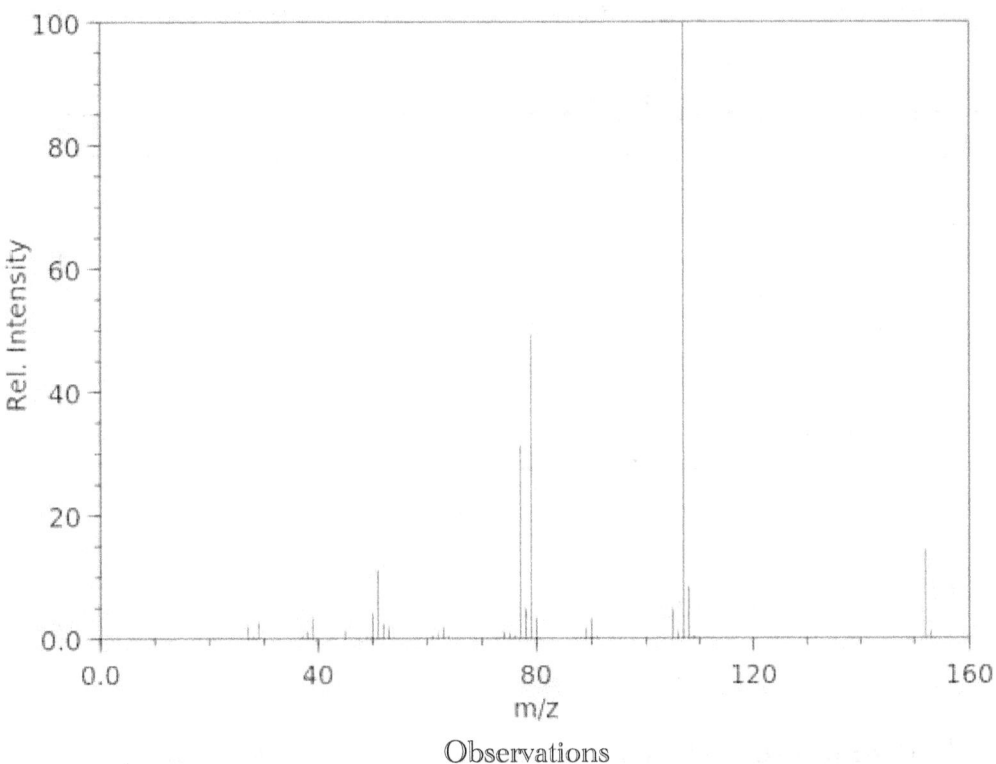

Observations

Charged fragments (m/z)	Assignment	Charged fragments (m/z)	Assignment
Molecular ion: 152	$[C_8H_8O_3]^+$	Base peak: 107	$[C_6H_5\text{-}CH_2O]^+$

Charged fragments (m/z)	Assignment	Charged fragments (m/z)	Assignment
89	$[C_6H_5C]^+$	89	$[C_6H_5 - C]^+$
79	$[C_6H_7]^+$	51	$[C_4H_3]^+$
77	$[C_6H_5]^+$	39	$[C_3H_3]^+$
74	$[C(OH)CO_2H]^+$	29	$[C_2H_5]^+$

Conclusions

- The peak at m/z = 77 indicates the presence of a benzene ring with one functional group attached;
- The difference between the molecular ion and the base peak is m / z = 45 which equates to $-CO_2H$;
- A very small peak at m/z = 74 suggests the existence of a carboxylic acid group with a hydroxyl group attached to one of the carbon atoms;
- The peak at m/z = 89 implies the existence of a monosubstituted benzene ring with the substituent being $- CH(OH)CO_2H$.

This means that the following possible candidate, which is consistent with the infrared spectrum, is plausible:

and we can determine whether or not this is correct by examining the ^1H and ^{13}C nmr spectra.

NMR Spectra

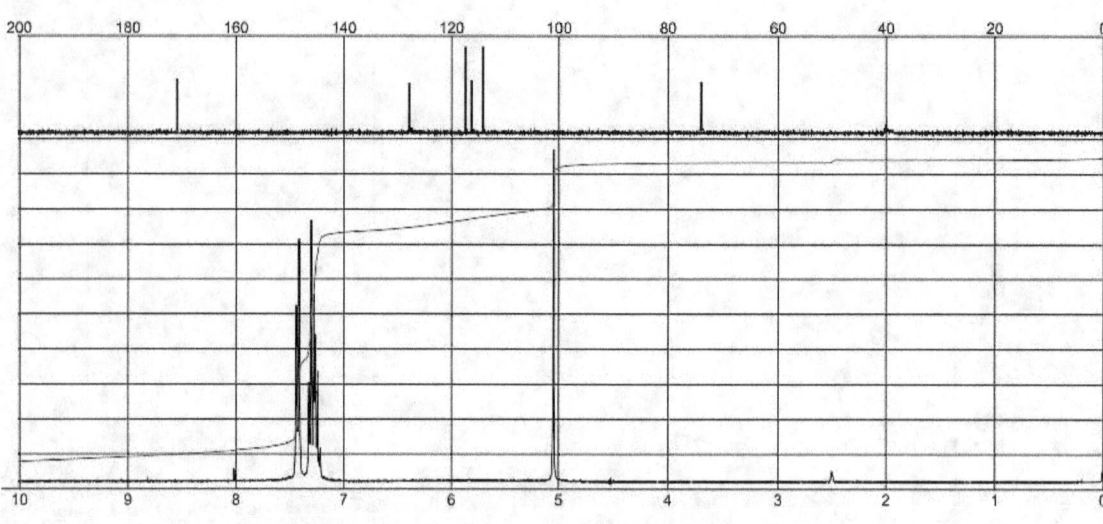

Observations and Conclusions

We will start with the ^{13}C nmr spectrum.

^{13}C NMR Spectrum

Chemical shift δ (ppm)	Integral	Assignment
172	1	C = O
128	1	Aromatic
118	2	Aromatic
117	1	Aromatic
115	2	Aromatic
75	1	C – O

Conclusions

This is straightforward to analyse and is interesting due to the presence of two pairs of chemically and magnetically equivalent aromatic carbon atoms.

We can readily assign the peaks at δ 172 ppm and δ 75 ppm as due to a carboxylic acid. The aromatic signals are more interesting though since there is only one possible general structure:

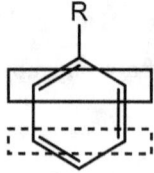

where R contains the residual $C_2H_2O_2$ atoms which must form a functional group.

The pairs of mutually chemically and magnetically carbon atoms are highlighted by the rectangles which account for the singlets of integral two at δ 118 and 115 ppm. At this stage it is not possible to assign the remaining two singlets of integral one to either of the remaining carbon atoms but this is the only plausible structure and we now need to consider the sole functional group, labelled as R in the above structure.

We know that there is a – CO_2H group and this then leaves us with fitting in the remaining C, H and O atoms. There are a number of ways in which carbon and oxygen can comprise part of a functional group such as an ether or an alcohol.

If the group contains an ether structure then we would have the following structure for R:

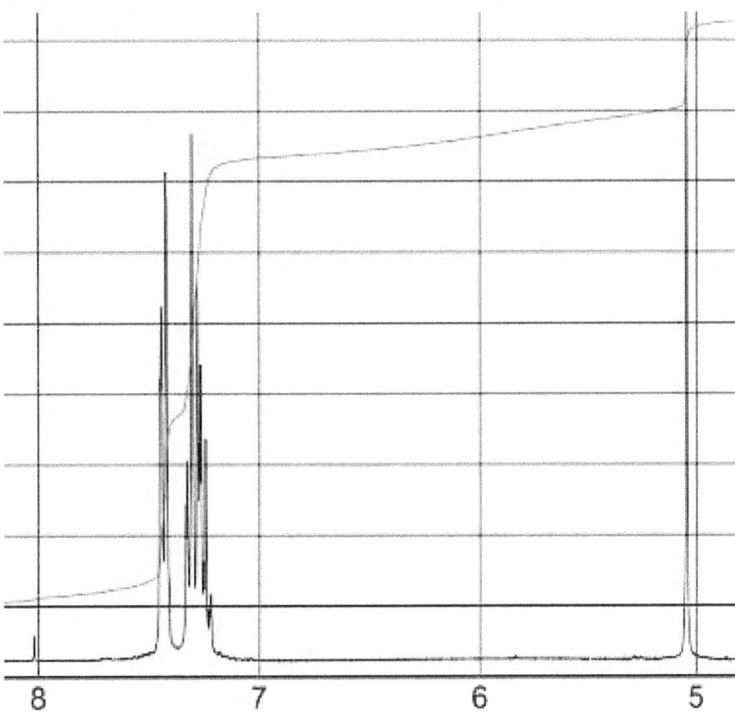

This cannot be correct since this molecule would have the formula $C_8H_7O_3$ and so there is an unassigned hydrogen atom. This leaves us with only one alternative:

This matches the molecular formula and the ^{13}C nmr spectrum and is also a good example of mixing different ways of drawing structures to the best effect. Although, confident of the structure which confirms the infrared and mass spectra, we must, of course, finally confirm the structure by examining the ^1H nmr spectrum.

^1H NMR Spectrum

This expanded spectrum is repeated below.

It is important to note that this spectrum, measured in D_2O (^1H_2O) does not show any signal for hydroxyl hydrogen atoms since they are labile and exchange with the deuterium from the D_2O so rapidly that they are not routinely detected. This is significant since the spectrum shows two signals in the integral ratio 5:1 reinforcing the suggestion that there are two – OH groups in the molecule.

There is too much overlap to fully analyse the aromatic spectrum but we can make the following assignments

Chemical shift δ (ppm)	Integral	Multiplicity	Assignment
7.50	2	Complex	b / b'
7.35	1	Complex	d
7.30	2	Complex	c / c'
5.10	1	Singlet	a

The rationale behind these assignments is as follows:

▪ The singlet due to – **HC** – O hydrogen atom is slightly outside the range in the data guide but it is only a guide.

▪ The multiplet of integral one can only be caused by the single hydrogen, H_d, as the other four hydrogen atoms form two pairs of hydrogens which are mutually chemically and magnetically equivalent *within* the pair.

▪ It is essential to try to distinguish between the two pairs of hydrogen atoms, H_b / $H_{b'}$ and H_c/$H_{c'}$.

  ▪ In its ^{1}H nmr spectrum, benzene displays a singlet at δ 7.30 ppm and this corresponds with the multiplet due to H_c/$H_{c'}$.

  ▪ This leaves H_b/$H_{b'}$ to be assigned to the multiplet centred on δ 7.50 ppm which will be deshielded due to the oxygen – containing substituent functional group.

Conclusions

Structure:

Systematic name: 2-Hydroxy-2-phenylacetic acid

Trivial name: Phenylglycolic acid

Chapter VII

Propofol

Discovered in 1977 and approved for use in 1989, ***propofol***, marketed as Diprivan, is a short – acting medication which causes decreased levels of consciousness and memory loss. This property makes it useful as a general anaesthetic and sedation. It is also used in assisted dying in Canada and was used as a sedative in executions in Missouri until exports for this purpose to America from the United Kingdom and European Union were prohibited.

At the time of writing (December 2022), propofol is used in combination with lipuro to sedate, through intravenous administration, Covid19 patients who are intubated in intensive care units.

It has largely replaced the traditional anaesthetic sodium thiopental as recovery from propofol is more rapid but it has a number of serious common adverse effects including an irregular heart rate, low blood pressure, a burning sensation at the site of injection and breathing interruption.

With a melting point of 18°C and a boiling point of 256°C, propofol is a colourless liquid which is easily solidified for ease of use and forms a milky emulsion when mixed with water due to its near complete insolubility in water.

Propofol has the elemental composition C: 80.77%, H: 10.20%, O: 8.97% and, with a formula mass of 178.28 g mol^{-1}, it has the **empirical** and **molecular** formulas $C_{12}H_{18}O$.

Infrared Spectrum

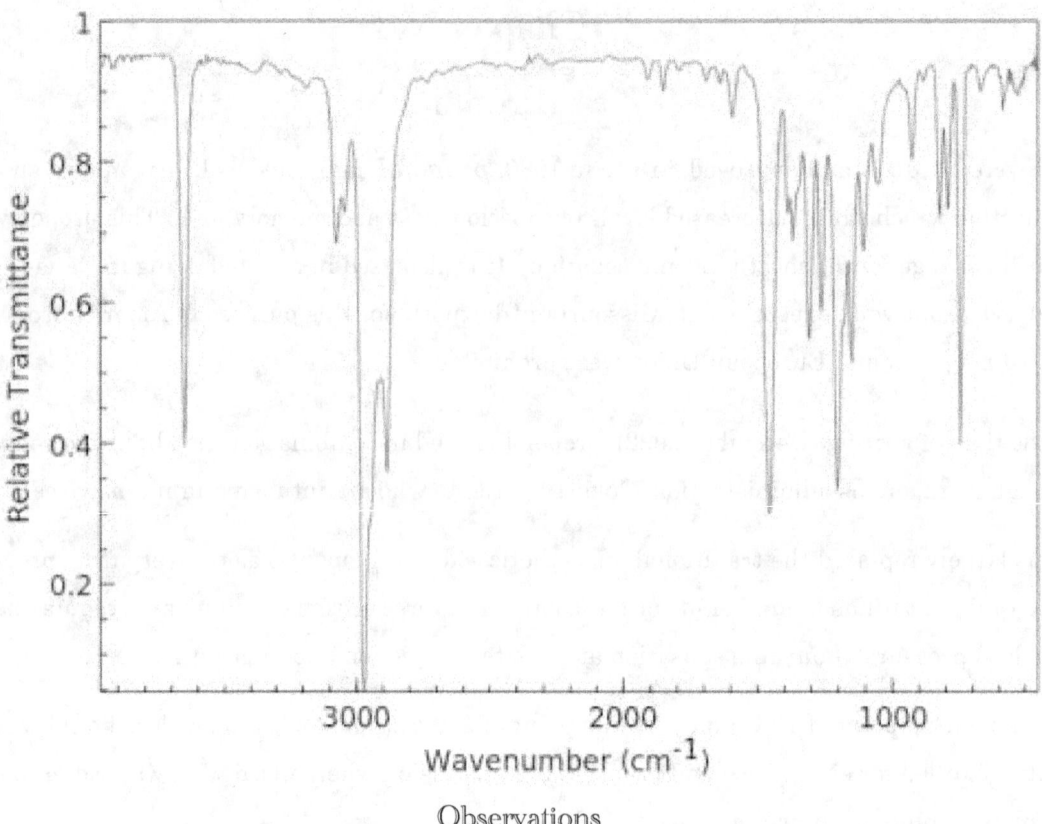

Observations

(√ / X)	Wavenumber range (cm⁻¹)	Wavenumber (cm⁻¹)	Assignment
√	3200 - 3700	3620	O – H
X	3200 - 3600		N – H
√	3000 – 3300	3090,3080	C – H (aromatic)
√	2500 – 3000	2990, 2820	C – H (aliphatic)
X	2200 – 2500		C ≡ N
X	1700 – 1800		C = O
X	1600 – 1700		C = C (aliphatic)
X	1585 – 1600		C – C (aromatic)
√	1450 – 1600	1520	C – C (aromatic)
√	1000 – 1300	1180	C – O
X	700 – 1000		C – X (X = Cl, Br or I)

Conclusions

This compound is aromatic, has aliphatic substituents and an – OH group. The aliphatic substituents cannot be alkenes as there is no evidence for a C = C bond. Essentially, therefore, the spectrum suggests that the molecule is a substituted phenol where the substituents are aliphatic.

If the molecule has only one alkyl substituent then, since the phenol molecule, itself, accounts for C_6H_4OH leaving C_6H_{13} to be accounted for. Similarly, if the phenolic ring has two alkyl substituents then the ring accounts for C_6H_3OH leaving C_6H_{14} to be accounted for. We can continue in a similar manner for rings with even more substituents and learn more from the mass spectrum.

Mass Spectrum

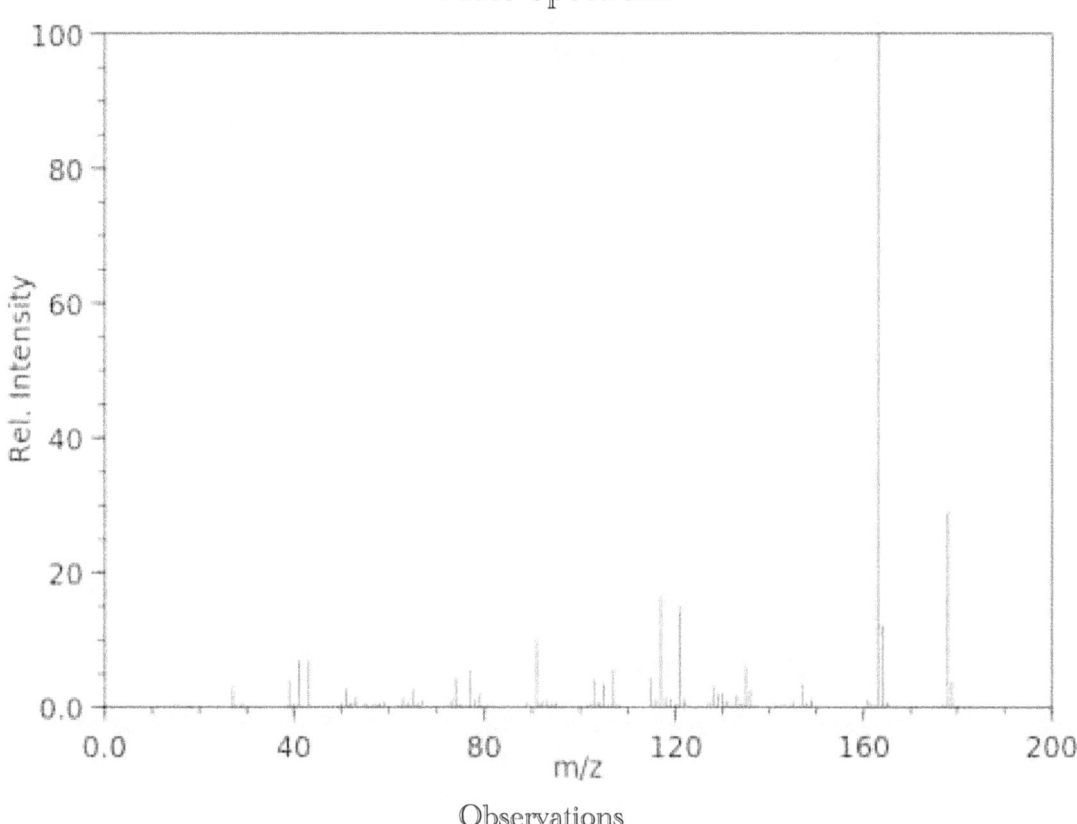

Observations

Charged fragments (m/z)	Assignment	Charged fragments (m/z)	Assignment
Molecular ion: 178	$[C_{12}H_{18}O]^+$	Base peak: 163	$[C_{11}H_{15}O]^+$

121	$[C_6H_5 - CH_2 - (CH_3)_2]^+$	77	$[C_6H_5]^+$
117	$[C_6H_5 - C (CH_2)_2]^+$	76	$[C_6H_4]^+$
91	$[C_6H_5 - CH_2]^+$	43	$[C_3H_7]^+$

Conclusions

⁕ There are a very large number of peaks separated by m/z = 1 which, of course, only be due to loss of hydrogen atoms. This is not surprising as there are eighteen hydrogen atoms in the molecular formula but it also makes it difficult to assign them since on their journey towards the detector individual molecular ions can fragment in different ways. In effect, the mass spectrum is a summary of all of the multiple and different fragmentations

⁕ The number of peaks around m/z = 76 supports the supposition that the molecule contains an aromatic ring although it implies that there is only one substituent on the ring which is impossible.

⁕ The peak at m/z = 43 indicates the presence of one or more isopropyl, $-C(H)(CH_3)_2$, groups.

⁕ The difference between the mass/charge ratios of the molecular ion and the base peak is m/z =15 which indicates the presence of a methyl, $-CH_3$, group which could arise from a methyl substituent directly bonded to the ring or from the fragmentation of the isopropyl group.

In the case of this molecule, the data guide is of relatively little help so we must rely on the ^1H and ^{13}C NMR spectra.

NMR Spectra

The ^1H and ^{13}C nmr spectra for propofol are displayed below:

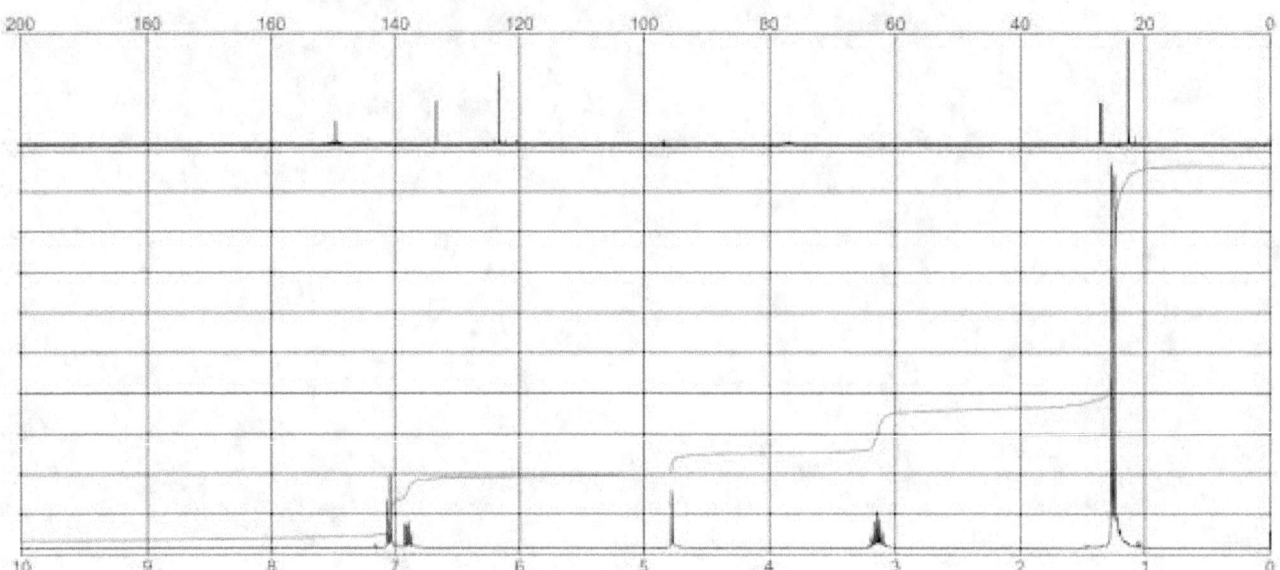

Both spectra are very clear and informative and we will start with the ^{13}C nmr spectrum.

^{13}C NMR Spectrum

The ^{13}C nmr spectrum contains the following peaks which can be assigned to some extent:

Chemical shift δ (ppm)	Integral	Assignment
150	1	Aromatic C
135	2	Aromatic C
124	3	Aromatic C
30	2	C – C
24	4	C – C

Quite clearly the molecule contains an aromatic ring and one peak, δ 150 ppm, is deshielded implying that this must have an electronegative atom bonded to it so the molecule is a substitued phenol.

The presence of four chemically and magnetically equivalent carbon and atoms and another two chemically and magnetically equivalent carbon atoms indicates that there must be two C – C(H)(CH$_3$)$_2$ groups attached to the ring and these can be assumed to be bonded to the two aromatic carbon atoms at δ 135 ppm.
This suggests that the structure is of one of two forms:

A   **B**

The two propyl groups in Isomer A will be deshielded to a greater extent than in Isomer B but the chemical shifts are inconclusive on this matter and so we now need to examine the ^1H nmr spectrum and predict the spectrum first.

Isomer A

There is a lot of symmetry in this molecule as shown below:

and it becomes even clearer if we label the hydrogen atoms as shown below:

We can predict that:

- The hydroxyl hydrogen atom, H_a, might produce a singlet but, being labile, may also not be observed.
- The two chemically and magnetically equivalent – CH groups, H_b and $H_{b'}$, will produce a septet of integral two due to the (n+1) rule
- Four methyl groups, labelled H_c, $H_{c'}$, $H_{c''}$, $H_{c'''}$, are all chemically and magnetically equivalent
  Theye will all be doublets due to splitting by the hydrogen atoms on the adjacent carbon atoms and, all being chemically and magnetically equivalent, will produce a doublet of integral twelve.
- The aromatic region is extremely important since:
  - H_d will produce a doublet of doublets, of integral one due to splitting by H_e and $H_{e'}$;
  - $H_{d'}$ will also produce a doublet of doublets of integral one due to sequential splitting by H_e and then H_d.

  These will appear at the same chemical shift and so there will be a doublet of doublets of integral two.
  - H_e produce a triplet, of integral one, due to splitting by H_d and $H_{d'}$.

In other words we would observe a doublet of doublets of integral two due to H_c and $H_{c'}$ and a triplet of integral one assignable to H_d.

To summarise, we can predict the signals and their multiplicities as shown below:

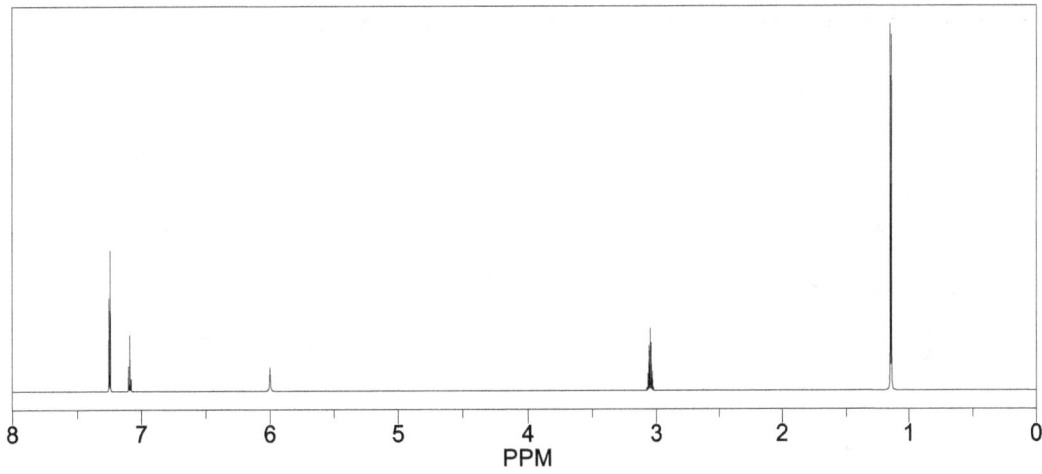

Isomer B

This molecule will also be symmetrical but there will be some significant differences. Using the labelling below:

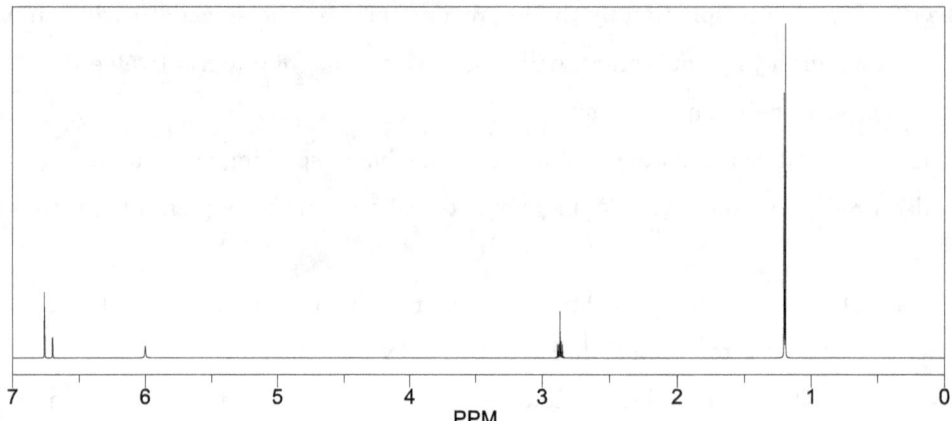

we can predict that:

* If it appears, H_f will produce a singlet of integral one;
* There will be a doublet of integral twelve due to H_j, $H_{j'}$, $H_{j''}$, $H_{j'''}$, all split by the adjacent – CH hydrogen atom, H_i and $H_{i'}$ respectively;
* H_i will produce a septet, of integral one, due to splitting by H_j and $H_{j'}$ whilst $H_{i'}$ will also produce a septet due to splitting by $H_{j''}$ and $H_{j'''}$. The cumulative effect will be a septet of integral two since H_i and $H_{i'}$ are chemically and magnetically equivalent and H_j, $H_{j'}$, $H_{j''}$ and $H_{j'''}$ are all mutually chemically and magnetically equivalent.
* The aromatic region is more significant since, in contrast to Isomer A, H_g, $H_{g'}$ and H_h will all produce singlets, of integral one, since none have hydrogen atoms on adjacent carbon atoms.

The predicted spectrum should be similar to this:

If we examine the observed ¹H nmr spectrum, it is clear that there are not three singlets of integral one so we can discount Isomer B as a candidate. In reality, there is a doublet of integral two and a triple of integral one and this reconciles with the predicted spectrum of Isomer A. The structure is confirmed by the spectrum.

Conclusions

Structure:

Systematic name: 2,6-bis(propan-2-yl)phenol

Chapter VIII

Paracetamol

Acetaminophen is used to treat fever and mild to moderate pain including migraines and is marketed as *paracetamol*.

It is fairly effective in reducing temperature but it is not as effective as butylphenylpropionic acid which is marketed in much of the world as nurofen, ibuprofen, advil and motrin.

With mild to moderate headaches and other pains, it has been found that the effectiveness of this compound is enhanced when taken in combination with aspirin and caffeine (a formulation which is readily available over the counter) and also in combination with aspirin and codeine. This formulation is available under a number of trade names including solpadeine and co-codamol.

It is also used to treat post-surgery pain since this compound has fewer side effects than aspirin but can cause nausea and abdominal pain. Large doses can be fatal due to causing multiple organ failure and it has also been found to aggravate asthma in pregnant women.

A colourless, crystalline solid with a melting and boiling points of 169°C and 420°C its elemental C:63.50 %, H: 6.01%, N: 9.27%, O: 21.17% and a formula mass of 151.17 g mol^{-1}.

This means that both the **empirical** and **molecular** formulas are $C_8H_9NO_2$.

Infrared Spectrum

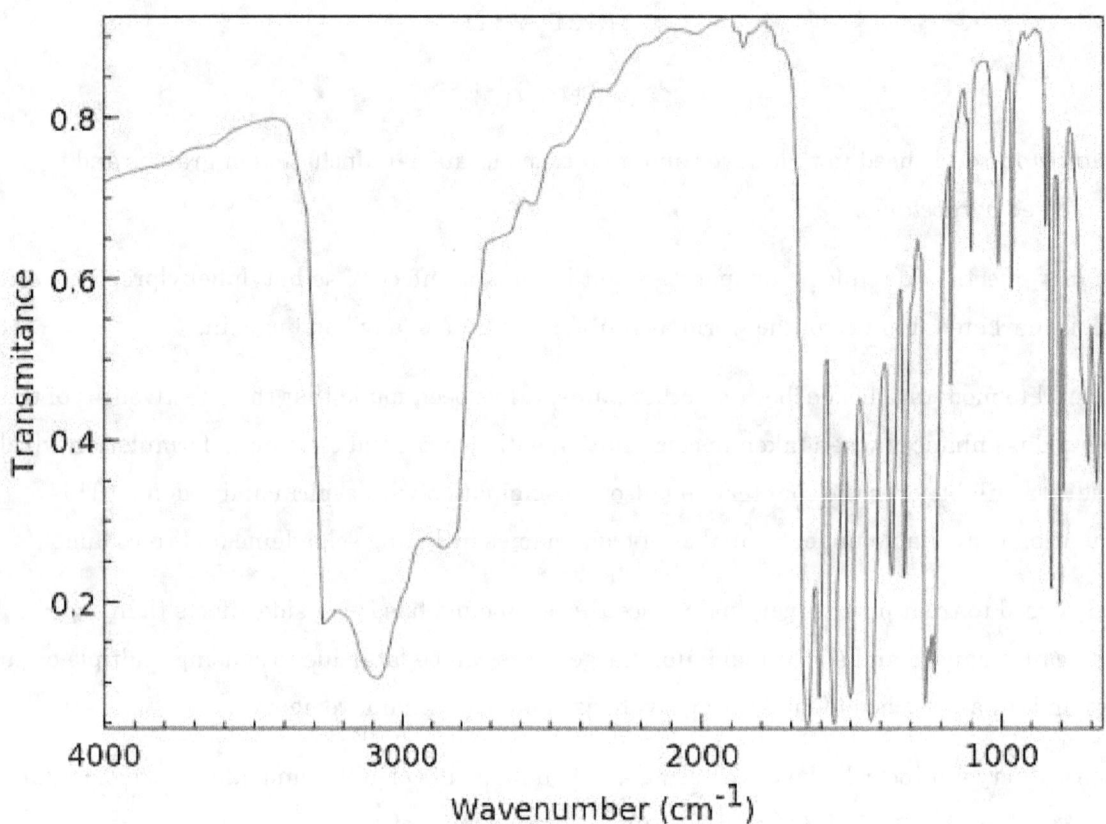

Observations

(√ / X)	Wavenumber range (cm⁻¹)	Wavenumber (cm⁻¹)	Assignment
√	3200 - 3700	3290	O – H
√	3200 - 3600	3290	N – H
√	3000 – 3300	3090	C – H (aromatic)
√	2500 – 3000	2800	C – H (aliphatic)
X	2200 – 2500		C ≡ N
√	1700 – 1800	1650	C = O
X	1600 – 1700		C = C (aliphatic)
X	1585 – 1600		C – C (aromatic)
X	1450 – 1600		C – C (aromatic)
√	1000 – 1300	1250	C – O
X	700 – 1000		C – X (X = Cl, Br or I)

Conclusions

This spectrum has some very broad peaks in the O – H and N – H, aromatic and aliphatic C – H regions.

▪ The peak assigned to C = O is slightly outside of the range specified in the data guide.

▪ Given the number of carbon atoms this molecule is likely to possess an aromatic ring and is possibly a substituted phenol with an amino and carboxylic acid group.

Mass Spectrum

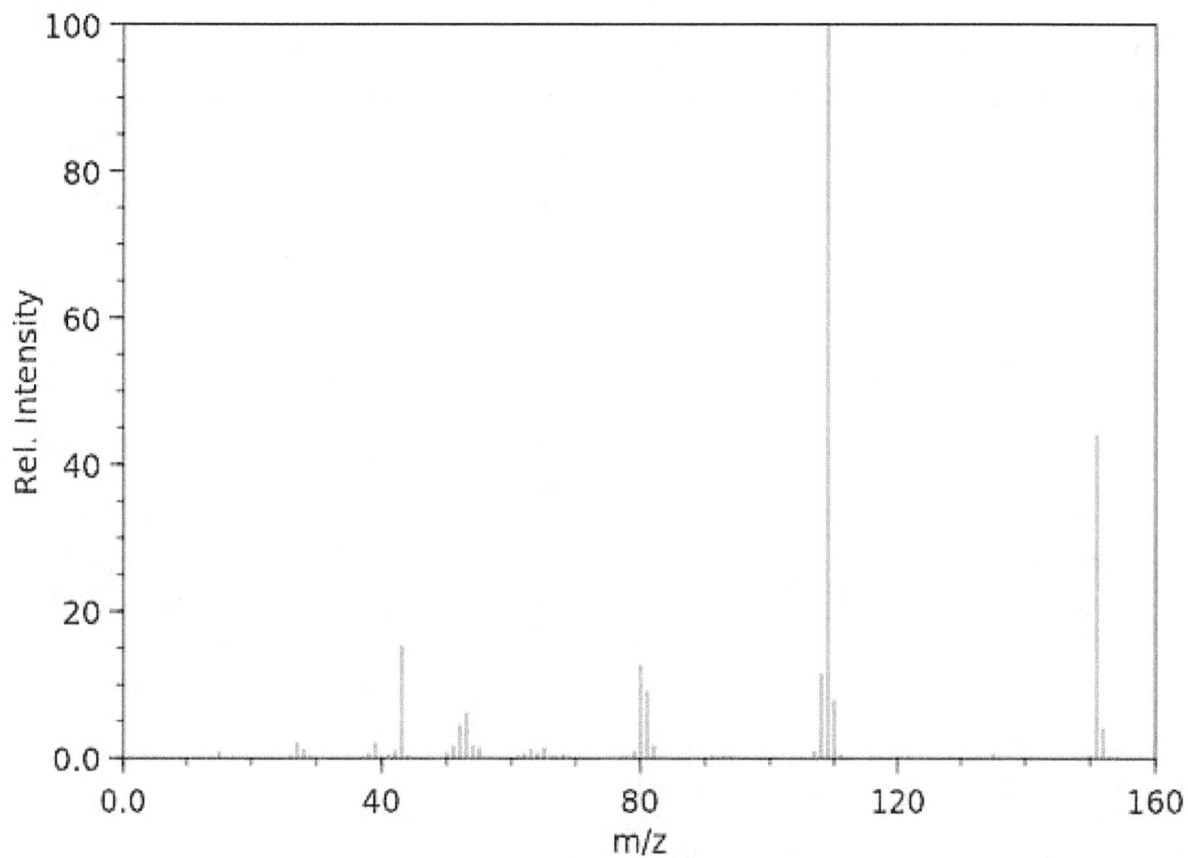

Observations

Charged fragments (m/z)	Assignment	Charged fragments (m/z)	Assignment
Molecular ion: 151	$[C_8H_9NO_2]^+$	Base peak: 109	$[C_7H_9O]^+$

135	$[C_8H_8NO]^+$	43	$[C_3H_7]^+ / [CH_3C{=}O]^+$
80	$[C_6H_8]^+$	39	$[C_3H_3]^+$

Conclusions

It is difficult to draw exact conclusions from the infrared and mass spectra but it is clear that the molecule contains a benzene ring but it is clear, from the infrared spectrum that the molecule contains:

- An – OH group;
- An ester grouping and;
- An N – H group.

It still seems reasonable to assume that the base molecule is:

where the squiggles represent, as yet, undetermined groups or may even just be hydrogen atoms.

There must be at least one other functional group to accommodate the remaining C_2H_4NO fragment. Given the limited number of hydrogen atoms it is unlikely that this cluster would form two or more additional functional groups so the next task is to assemble them into a feasible fragment. From the infrared spectrum, we already know that there is a C=O group and an N – H bond and so a possible functional group is:

where the dashed line indicates the bond between the group and the benzene ring.

This seems promising and there are then five possible isomers as shown below:

If we examine the structures closely we observe that:-

- Isomers I and V are mirror images of each other as are;
- Isomers II and IV.

We can now predict the ^1H and ^{13}C nmr spectra to identify which, if any isomer, is correct.

This means that we need only consider isomers I, II and III and all these fit the observed stretches in the infrared spectrum and justify the significant peaks due to all the significant indicated in the infrared table by a tick.

Examining the significant peaks in the mass spectrum, we can make a different assignment to the m/z = 43 peak which was, initially, assigned to $[C_3H_6]^+$ and / or $[COCH_3]^+$ both of which are consistent with all three of the candidate isomers.

There is also a peak at m/z = 15 which can only be assigned to a methyl, –CH3, group and this is also consistent with all three proposed candidates.

Finally, with regard to the mass spectrum, the difference between the molecular and base ions, m/z = 42, this can be assigned to $[CH_2C=O]^+$ which is also consistent with all three proposed structures.

We can make many other possible assignment but, sometimes, it can be futile as we have no way of proving if they are correct so our next task to examine the NMR spectra in order to identify the correct structure of paracetamol.

NMR Spectra

The ^1H and ^{13}C nmr spectra are displayed below and we will examine the ^1H nmr spectrum first and start by predicting the spectra for all three candidate structures before examining the ^1H nmr spectrum in detail.

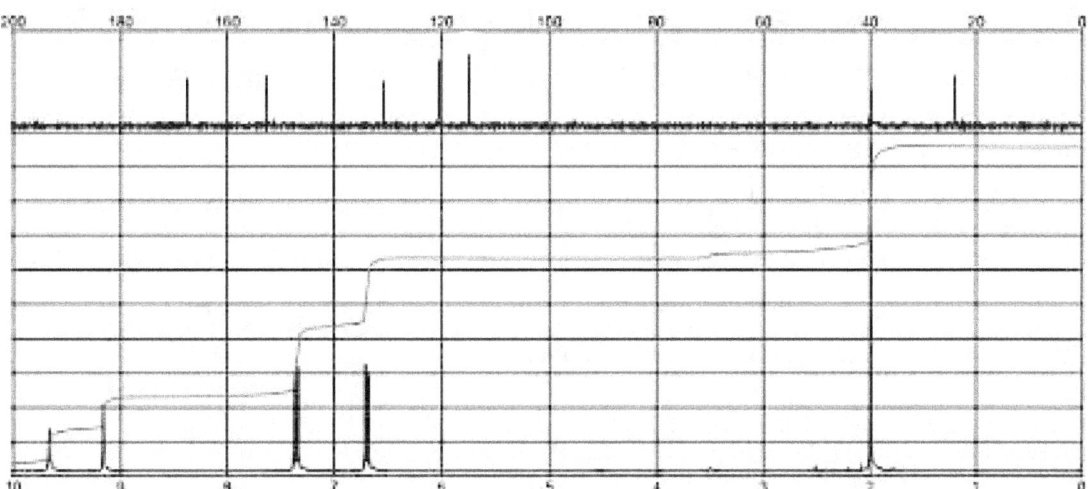

^1H NMR Spectrum

All the isomers will produce signals due to the substituents to the ring which will be extremely similar but the aromatic hydrogen atoms will be different in all three isomers as shown below in the labelled diagrams.

- **Isomer I** will produce four distinct signals, all of integral, one due to the hydrogen atoms labelled, a,b,c,d.
    - H_a will produce a doublet of doublets of doublets due to sequential splitting by H_b, H_c and then H_d.
    - H_b will produce a doublet of triplets since it will be split by H_a and H_c (n+1 rule) which will be split into a doublet of triplets by H_d.
    - H_c will produce a doublet of triplets since it will be split by H_b and H_d (n+1 rule) which will be split into a doublet of triplets by H_a.
    - H_d will produce a doublet of doublets of doublets due to sequential splitting by H_c, H_b and then H_a.

- **Isomer II** will also produce four distinct signals, all of integral one, due to the hydrogen atoms labelled e,f,g,h.
    - H_e will produce a doublet of doublets due to its original singlet being split into a doublet by H_f and this doublet will be split into a doublet of doublets by H_g.
    - H_f will produce a triplet since its signal is split by H_e and H_g.
    - Hg will produce a doublet of doublets due to its original singlet being split into a doublet by H_f and this doublet will be split into a doublet of doublets by H_e.
    - Since there are no hydrogen atoms on the adjacent hydrogen atoms, Hh will produce a singlet.

* **Isomer III** has two pairs of chemically and magnetically hydrogen atoms, H_i / $H_{i'}$ and H_j / $H_{j'}$, respectively.

   * H_i and $H_{i'}$ will each produce a doublet since each has one hydrogen (the other of this pair) on immediately adjacent carbon atoms. Since H_i and $H_{i'}$ are chemically and magnetically equivalent they will resonate at the same frequency, giving rise to a doublet of integral **two**.

   * Likewise, H_j and $H_{j'}$ will each produce a doublet since each has one hydrogen (the other of this pair) on immediately adjacent carbon atoms. Again, since they are mutually chemically and magnetically equivalent they will resonate at the same frequency, also giving rise to a doublet of integral **two**.

In summary, the aromatic region of isomer,

   * I will contain two doublets of doublets of doublets and two doublets of triplets all of integral **one**;

   * II will produce a singlet, two doublets of doublets and a triplet all of integral **one**.

   * III will give rise to two doublets each of integral **two**.

If we examine the aromatic region of the ^1H NMR spectrum we observe two doublets of integral two and so it is clear that only **Isomer III** matches this spectrum.

Examining the remainder of the spectrum, the methyl signal is clear and the only discussion remaining is the assignment of the peaks at δ 9.65 and δ 9.20 ppm. One will be due to the – OH hydrogen and the other to the – NH hydrogen atom.

Since oxygen is more electronegative than nitrogen this means that the – OH hydrogen will be more deshielded i.e. exposed and so we can assign this hydrogen atom to δ 9.65 ppm.

We also need to determine the assignments of H_i / $H_{i'}$ and H_j / $H_{j'}$ and we can apply the same principles of electronegativity. Since H_i / $H_{i'}$ are close to the more electronegative halogen atom it is fair to assume that these are more deshielded than H_j / $H_{j'}$ which are close to the oxygen atom.

We can, therefore, confidently, make the following assignments:

Chemical shift δ (ppm)	Integral	Multiplicity	Assignment
9.65	1	Singlet	– OH
9.20	1	Singlet	– NH
7.40	2	Doublet	H_i / $H_{i'}$
6.75	2	Doublet	H_j / $H_{j'}$
2.00	3	Singlet	– CH_3

Naturally we must confirm these assignments and this structure by examining the ^{13}C nmr spectrum and this is our last task.

^{13}C NMR Spectrum

We can label the carbon atoms as shown below:

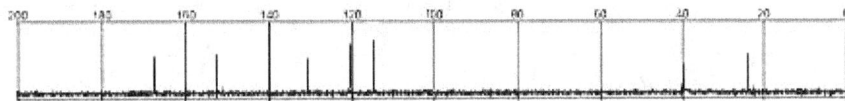

- As with the ^1H nmr spectrum the four unsubstituted carbon atoms form two pairs of mutually chemically and magnetically equivalent atoms, C_n / $C_{n'}$ and C_o / $C_{o'}$ respectively and there will be two other aromatic carbon atoms C_m and C_p. This means that we will have *two* singlets of integral **two** and *two* singlets of integral **one** in the aromatic region (δ 110 – 160 ppm);
- There will be a singlet due to C_q in the C = O region (δ 160 – 220 ppm) and;
- There will a singlet due C_r in the alkyl region (δ 0 – 50 ppm).

If we examine the actual ^{13}C nmr spectrum which is repeated below, we observe that this prediction matches the observed spectrum:

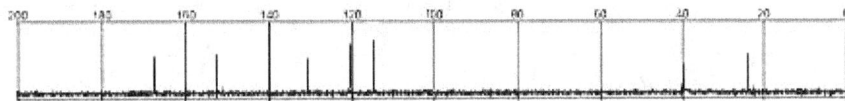

It is clear that:-

- The C = O carbon atom resonates at δ 168 ppm and;
- That the methyl, – CH_3, carbon atom resonates at δ 23 ppm.

Again, we can consider the electronegativity of the substituent halogen atoms.

- The two pairs of carbon atoms can be distinguished since C_m / $C_{m'}$ is closer to the more electronegative, oxygen, atom so will be deshielded more than C_n / $C_{n'}$ which are closer to the slightly less electronegative, nitrogen, atom and will be the cause of the doublet of integral two at δ 121 ppm whilst the slightly less deshielded C_n / $C_{n'}$ pair will give rise to the singlet of integral two at δ 115 ppm.

- This leaves us with assigning the peaks, both of integral one, at δ 154 and δ 132 ppm. One of the carbon atoms, C_m, has the – OH bonded to it whilst the other, C_p, has the nitrogen atom bonded to it. Since oxygen is more electronegative than nitrogen we can assign the δ 154 ppm to C_m and the peak at δ 132 ppm to C_p.

Conclusions

Since this is quite a lengthy discussion the assignments can be summarised with the labelling as shown below:

For the ^{1}H nmr spectrum   For the ^{13}C nmr spectrum

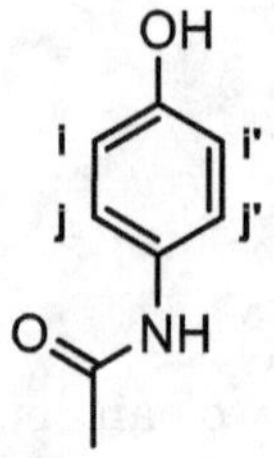

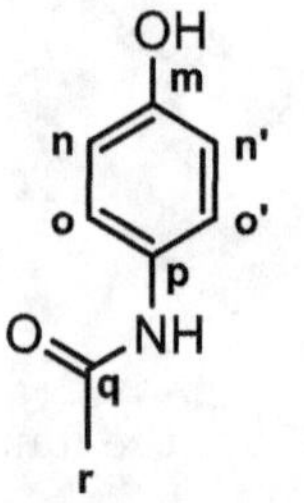

^{1}H NMR Spectrum

Chemical shift δ (ppm)	Integral	Multiplicity	Assignment
9.65	1	Singlet	– OH
9.20	1	Singlet	– NH
7.40	2	Doublet	H_i / $H_{i'}$
6.75	2	Doublet	H_j / $H_{j'}$
2.00	3	Singlet	– CH_3

^{13}C NMR Spectrum

Chemical shift δ (ppm)	Integral	Assignment
168	1	C_q
154	1	C_m
132	1	C_p
121	2	C_n / $C_{n'}$
115	2	C_o / $C_{o'}$
23	1	C_r

Conclusions

Structure:

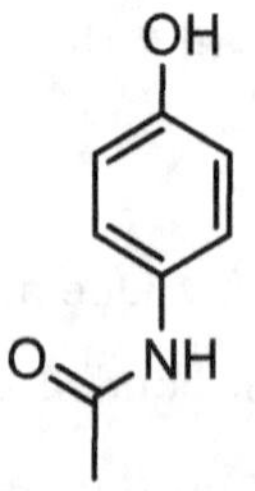

Systematic name: N-(4-hydroxyphenyl)acetamide

Trivial names: N-acetyl-para-aminophenol

Trade names: Paracetamol, Tylenol, Calpol (as a syrup for children), Panadol and numerous others, often country of sale – specific.

Chapter IX

Ferulic acid

Ferulic acid is a major constituent of fruits and vegetables, particularly wheat, oats and rice but it also found in apples, peanuts and coffee. It has been found to have strong anti – oxidant and anti – inflammatory properties. At a concentration of 0.5% (v/v) it is believed to inhibit melanin formation and is used in sun screen creams.

Due to its anti-oxidant properties, the compound is used in some cancer treatments since it is reactive towards free radicals and, for the same reason, it is also widely used in cosmetics especially anti-ageing products. It is also used as a diabetes treatment since hyperglycaemia causes excessive production of free radicals.

Ferulic acid has been identified in a number of herbs used in Chinese medicines such as ginseng and has also been found in green tea. It is also found in some herbal infusions, brewed from the European centaury flowering plant which is used a medicinal herb in many parts of Europe, as a treatment for gastric and liver disease.

With a melting point of 170°C and a boiling point of 372°C ferulic acid is, at room temperature, a colourless crystalline solid and is sparingly soluble in cold water

Ferulic acid has the elemental composition, C: 61.80%, H: 5.20%, O: 32.96% and has a formula mass of 194.18 g mol^{-1}.

This means that the **empirical** formula is $C_5H_5O_2$ whilst its **molecular** formula is $C_{10}H_{10}O_4$.

Infrared Spectrum

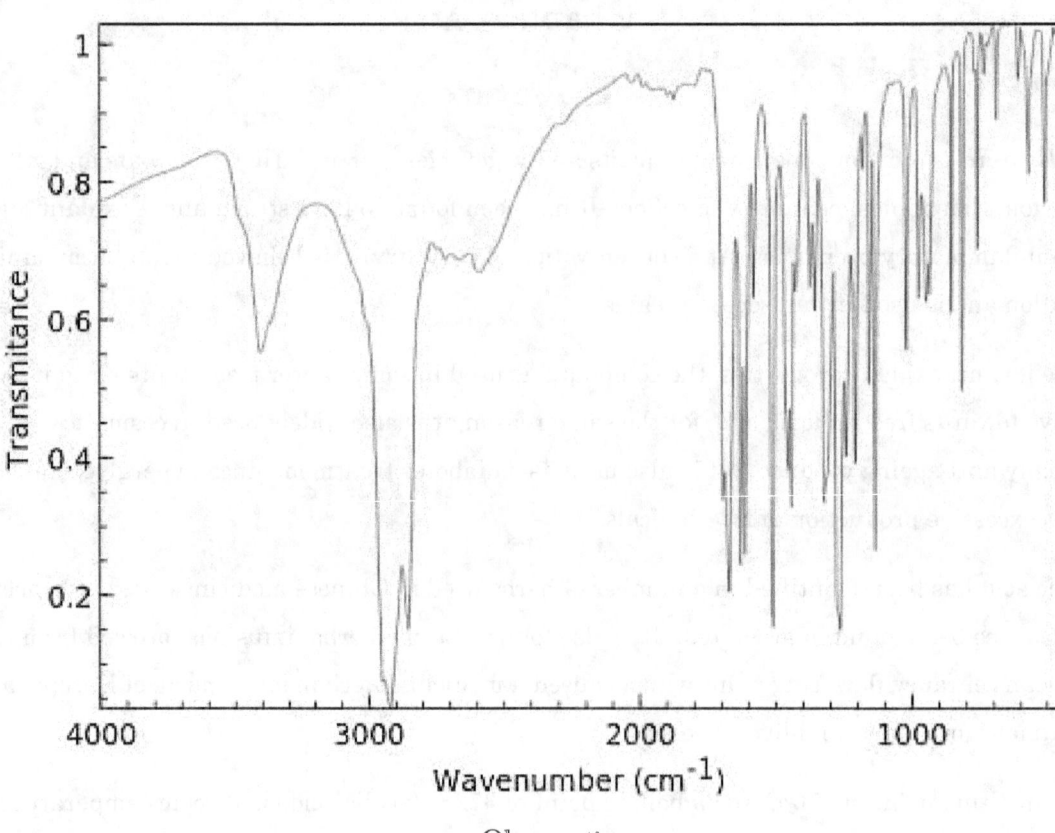

Observations

(√ / X)	Wavenumber range (cm⁻¹)	Wavenumber (cm⁻¹)	Assignment
√	3200 - 3700	3400	O – H
X	3200 - 3600		N – H
X	3000 – 3300		C – H (aromatic)
√	2500 – 3000	2950, 2920, 2820	C – H (aliphatic)
X	2200 – 2500		C ≡ N
√	1700 – 1800	1720	C = O
√	1600 – 1700	1650	C = C (aliphatic)
X	1585 – 1600		C – C (aromatic)
√	1450 – 1600	1450, 1460, 1500	C – C (aromatic)
√	1000 – 1300	1280	C – O
X	700 – 1000		C – X (X = Cl, Br or I)

Conclusions

This compound contains a hydroxyl group and is also a carboxylic acid.

It contains a C = C bond and whilst the C – H stretches appear in the aliphatic region the 1450 – 1600 cm⁻¹ region indicate the presence of an aromatic ring. This makes sense as the molecular formula $C_{10}H_{10}O_4$ does contain sufficient hydrogen atoms for the molecule to be solely aliphatic.

Mass Spectrum

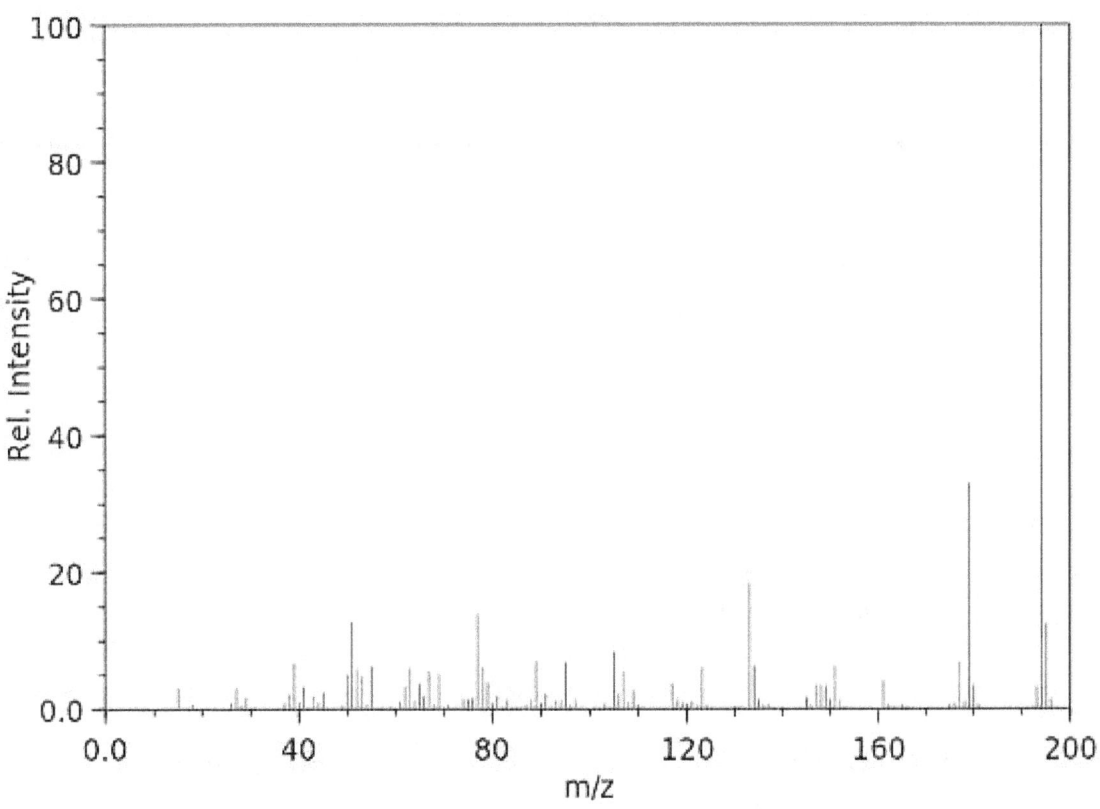

Observations

Charged fragments (m/z)	Assignment	Charged fragments (m/z)	Assignment
Molecular ion: 194	$[C_{10}H_{10}O_4]^+$	Base peak: 194	$[C_{10}H_{10}O_4]^+$

Charged fragments (m/z)	Assignment	Charged fragments (m/z)	Assignment
179	$[C_9H_{10}O_4]^+$	105	$[C_8H_9]^+$ / $[C_7H_5O]^+$
161	$[C_8HO_4]^+$ / $[C_9H_5O_3]+$	77	$[C_6H_5]^+$
133	$[C_8H_5O_2]^+$	51	$[C_4H_3]^+$

Conclusions

The most important peak in this spectrum is that at m/z = 77 which is only ever found to be caused by a benzene ring with *one* substituent and is also consistent with the infrared peaks at 1450, 1460 and 1500 cm^{-1} assignable to aromatic C – C bonds.

This can also be justified as there are insufficient hydrogen atoms for the molecule to be entirely aliphatic.

In the infrared spectrum there is a peak assignable to an O – H group and it might appear that a reasonable proposal is that this molecule is a mono-substituted phenol but this cannot be correct as there are other atoms to fit in and so it must due to some other entity.

There are simply too many peaks of low intensity to be confident about any assignments and so we must turn to the ^{1}H and ^{13}C nmr spectra which we consider next.

NMR Spectra

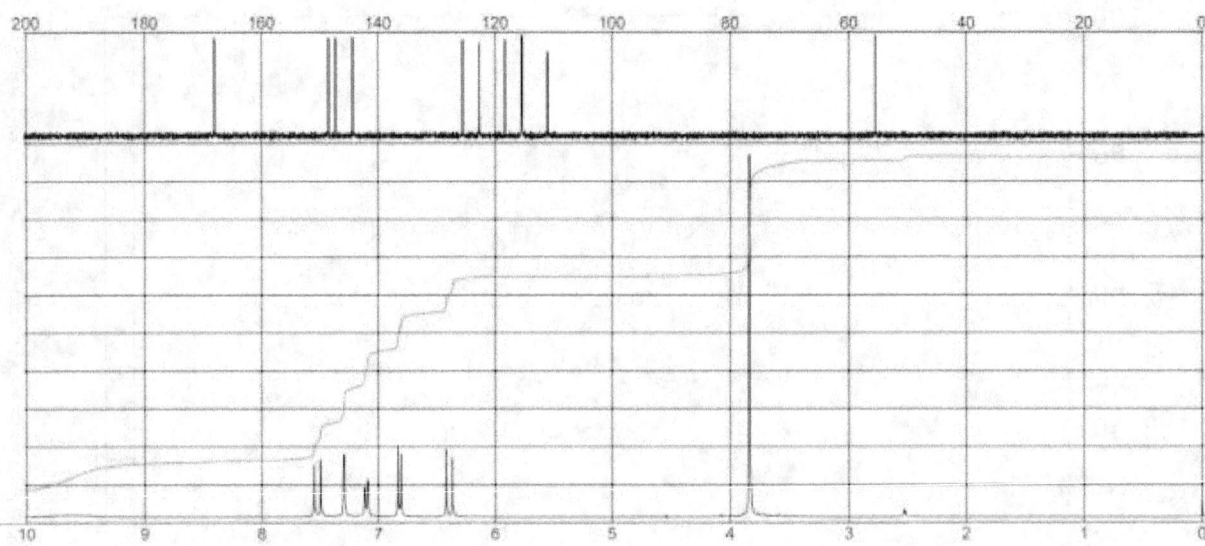

We will consider the ^{13}C nmr spectrum first.

^{13}C NMR Spectrum

Chemical shift	Integral	Assignment
169	1	C = O
150	1	Aromatic carbon*
148	1	Aromatic carbon*
145	1	Aromatic carbon*
127	1	Aromatic carbon*
123	1	Aromatic carbon*
118	1	Aromatic carbon*
115	1	C = C*
56	1	C – O

*There are seven signals in the aromatic / C = C region (δ 160 – 110 ppm)

Observations and Conclusions

- Six of the seven signals in the δ 160 – 110 ppm region are clearly due to the presence of a benzene ring and the seventh must be due to one carbon in a C = C bond. This is consistent with the presence of a C = C stretch in the infrared spectrum.

- The presence of only one C = C signal in this region implies that the other carbon in that bond is either shielded or deshielded, otherwise there would be two in approximately the same position.

- The existence of signals due to C – O and C = O indicate the presence of a carboxylic acid but that cannot be bonded directly to the ring and must form part of a chain substituent.

- There is also a C – O bond present somewhere in the molecule and an – OH group.

As is often the case, one option is to start with an educated guess, also known as a stab in the dark, or better still to examine the ^{1}H nmr spectrum.

IX – Ferulic acid

The expanded aromatic (δ 6 – 8 ppm) region is shown below:

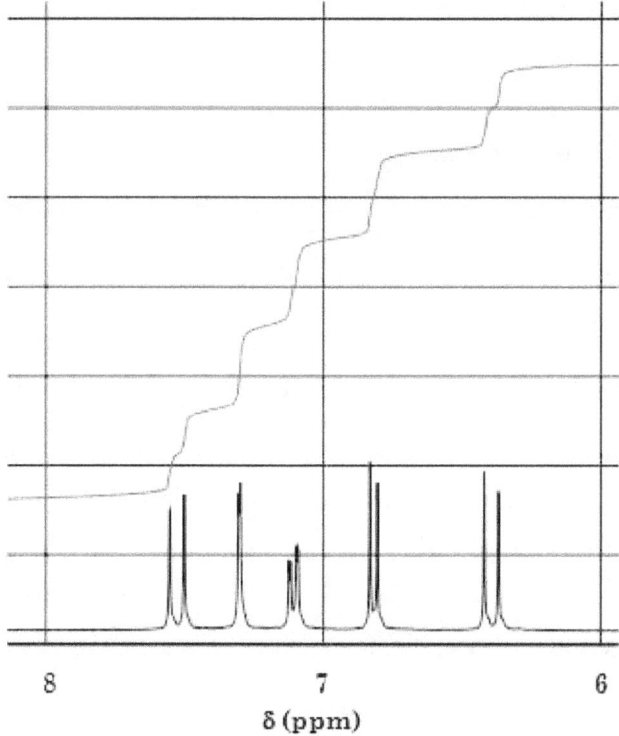

δ (ppm)

This region comprises signals due to the aromatic ring hydrogen atoms and the alkene hydrogen atoms.

- Alkene hydrogen atoms will produce signals of the same coupling constant (J) which is measured in Hertz (Hz). From the spectrum, there are two doublets with the same coupling constant J = Hz at δ 6.75 ppm and δ 7.1 ppm and these must be due to the two alkene hydrogen atoms.

- This leaves us with three signals at δ 7.3 ppm, δ 7.65 ppm and δ 7.70 ppm. The signals at δ 7.30 ppm and δ 7.70 ppm are doublets whilst that at δ 7.65 ppm is a singlet.

  If we examine the doublets then we can observe that the coupling constants are the same and this indicates that they are bonded to adjacent carbon atoms. The presence of a singlet indicates that there cannot be any hydrogen atoms on adjacent carbon atoms.

This indicates that the base molecule has one of two possible structures where R, R_1 and R_2 are, at this point, undefined:

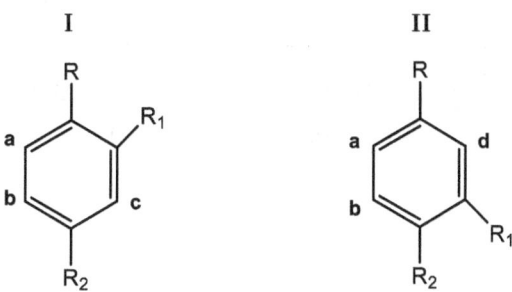

In **isomer I**,

- H_a will be split by H_b to form a doublet;
- H_b will be split by H_a to form a doublet whilst;
- H_c will be a singlet

and all signals will be of integral one.

50

In isomer II,

- H_a will be split by H_b to form a doublet;
- H_b will be split by H_a to form a doublet whilst;
- H_d will be a singlet

and, again, all signals will be of integral one.

This means that, in both cases, the signal due to H_c (isomer I) or H_d (isomer II) must be that observed at δ 7.65 ppm. This is significantly shifted from the δ 7.35 ppm signal observed in benzene and implies that there is at least one oxygen atom bonded to an adjacent carbon atom.

This leaves us to establish which of the signals, δ 7.70 ppm and δ 7.30 ppm, can be assigned to H_a and which can be assigned to H_b.

- The doublet at δ 7.30 ppm is extremely close to the singlet observed in benzene and this indicates that there are no electronegative (electron withdrawing) groups or electron donating groups on adjacent carbon atoms. The deshielded peak, at δ 7.70 ppm indicates that one of the adjacent carbon atoms must have an electron withdrawing atom attached and this can only be an – OH group simply because there are no other possibilities.

- The singlet at δ 7.65 ppm also indicates deshielding. This indicates that an adjacent carbon atom also has an oxygen atom attached.

- The singlet at δ 3.85 ppm can only be due to a – OCH_3.

This implies the following possible base structures:

where R equates to $C_3H_4O_2$.

We know from the infrared spectrum that the molecule also contains a carboxylic acid functional group and this can only occur at the end of a chain. This group accounts for a further CO_2H leaving us to assign the remaining C_2H_2 group which must be due to the alkene functional group which must be within the chain and this means that we have accounted for all the atoms but we still need to determine which of the following isomers is correct:

In isomer II, the highlighted aromatic hydrogen signals (below) will be deshielded to a greater extent and the signals due to the two highlighted hydrogen atoms will be close together.

This is not observed and so the structure must be **isomer I** which is wholly consistent with infrared, mass, ^{1}H and ^{13}C nmr spectra but we can go further in two respects: we can investigate the coupling constants in the hydrogen signals in the benzene ring and in the alkene bond.

From the data sheet we have the following:

Designation	Ortho –	Meta –	Para –
Structural formula			
Coupling constant range (J):	7 – 10	2 – 3	0 – 2

If we examine the determined structure again, with newly labelled hydrogen atoms, we have the following:

- H_a is ortho- to H_b and para- to H_c
- H_b is ortho- to H_a and meta- to H_c
- H_b is meta- to H_b and para- to H_a.

There are coupling constants assignable to all three orientations but the most important coupling constant is that present in the C = C bond. Again, from the data sheet, we have the following information about the coupling constants in C = C bonds:

Isomerism	Coupling constant (J) range (Hz)
Geminal	0 – 5
Vicinal (cis) / Vicinal (Z –)	5 – 14
Vicinal (trans) / Vicinal (E –)	15 – 20

The coupling constant in the alkene is J = 18 Hz and so the stereochemistry is trans, or in more modern terminology Entgegen (E-). We can finally confirm this structure and the conclusions are summarised next.

Conclusions

Structure:

Systematic name: (2E)-3-(4-hydroxy-3-methoxyphenyl)prop-2-enoic acid

Trivial name: coniferic acid.

Chapter X

Benzyl benzoate

First studied medically in 1918, ***benzyl benzoate*** is used as a treatment for scabies and lice and as a component of some asthma and whooping cough drugs. It is also used as an insect repellant particularly for ticks and mosquitoes.

It has few side effects beyond minor skin irritation but is not used with children and, although also used as a veterinary treatment, it is toxic to cats.

Industrially and in laboratory situations it used as a fixative in perfumes, polymer plasticisers and as a solvent for some derivatives of cellulose.

With a melting point of 18°C and a boiling point of 324°C, benzyl benzoate is a viscous, colourless liquid with a slight sweet balsamic odour at room temperature which forms colourless crystalline flakes with only slight cooling making it easy to handle.

It is readily synthesised by the action of benzyl chloride on sodium benzoate, even in aqueous solution, and even more conveniently by the action of sodium alkoxide on benzaldehyde.

It is also found naturally in the blossoms of hyacinths and tuberose which, a popular ornamental plant is a native of Mexico. It is a component of *Balsam of Peru* which is used in numerous commercial products including air fresheners, animal repellents, cleaning products, essential oils and as a food flavouring.

Benzyl benzoate is also found in tolu balsam which is used in some cough medicines, perfumes and as a natural remedy for skin conditions including rashes and dermatitis.

With a formula mass of 212.26 g mol^{-1} and elemental composition: C:79.15 %, H:5.71%, O:15.08% it has the **empirical** formula C_7H_6O and **molecular** formula $C_{14}H_{12}O_2$.

X – Benzyl benzoate

Infrared Spectrum

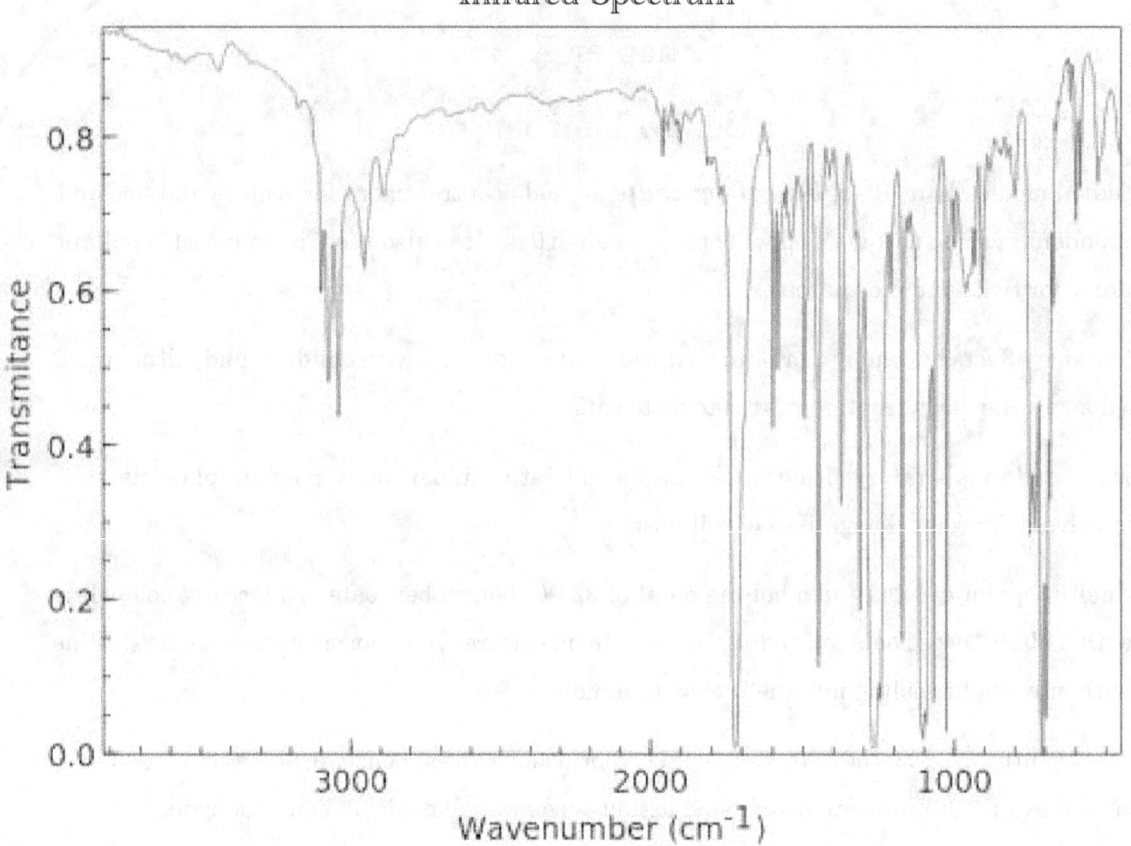

Observations

(√ / X)	Wavenumber range (cm⁻¹)	Wavenumber (cm⁻¹)	Assignment
X	3200 - 3700		O – H
X	3200 - 3600		N – H
√	3000 – 3300	3120,3080,3040	C – H (aromatic)
√	2500 – 3000	2980	C – H (aliphatic)
X	2200 – 2500		C ≡ N
√	1700 – 1800	1710	C = O
X	1600 – 1700		C = C (aliphatic)
X	1585 – 1600		C – C (aromatic)
X	1450 – 1600		C – C (aromatic)
√	1000 – 1300	1100 or 1280?	C – O
X	700 – 1000		C – X (X = Cl, Br or I)

Conclusions

- This molecule is aromatic and contains an ester grouping as shown by the presence of the C = O and the C – O stretch. The assignments to the latter bond are tentative as there are so many peaks in the fingerprint reason.

- A molecule containing an ester functional group will also show a signal assignable to C – O but this will appear in the 1300 – 1000 cm⁻¹ which is extremely cluttered even for a fingerprint region.

Mass Spectrum

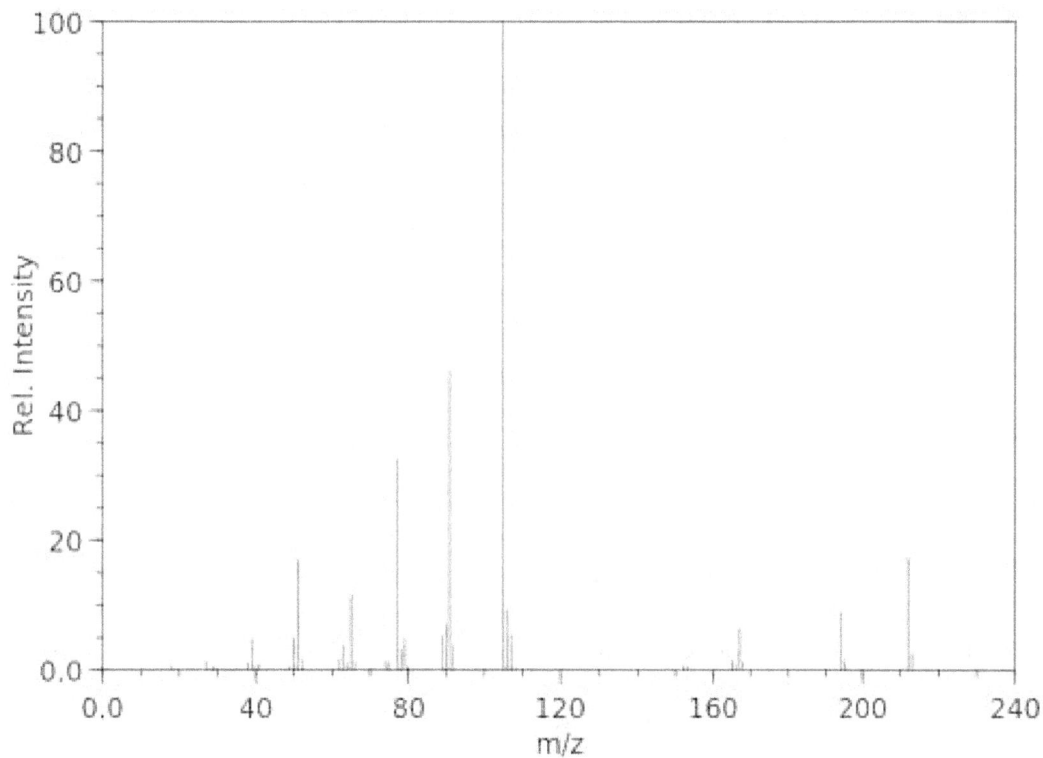

Observations

Charged fragments (m/z)	Assignment	Charged fragments (m/z)	Assignment
Molecular ion: 212	$[C_{14}H_{12}O_2]^+$	Base peak: 105	$[C_6H_5 - C = O]^+$
194	$[C_{14}H_{10}O]^+$	77	$[C_6H_5]^+$
167	$[C_{12}H_7O]^+$	65	$[C_5H_5]^+$
91	$[C_6H_5 - CH_2]^+$	51	$[C_4H_3]^+$

Conclusions

⊞ Due to the peak at m/z = 77, this molecule contains a mono – substituted benzene ring;

⊞ The base peak, m/z = 105, indicates that the substituent begins with a ester linkage – this is suggested by the presence of both C = O and C – O bonds whilst the absence of an O – H stretch shows it cannot be a carboxylic acid which, in any event would be a terminal group and could not create a linkage.

⊞ The presence of a peak at m/z = 91 indicates that there is an aromatic ring with a – CH₂ – substituent. This cannot be on the same aromatic ring and hence implies that the molecule comprises two, mono – substituted, aromatic rings linked by an ester grouping.

This all suggests that the following structure is plausible

but we can learn much more from the ¹H and ¹³C NMR spectra or start again.

NMR Spectra

The ^{1}H and ^{13}C nmr spectra are displayed below:

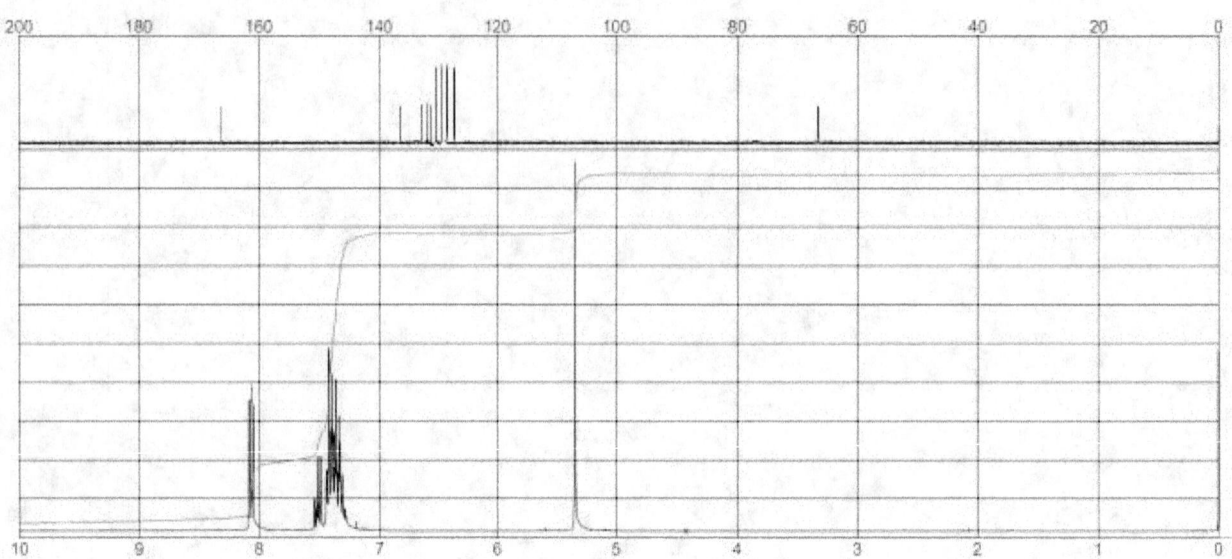

The ^{1}H nmr spectrum is quite daunting so it is preferable to consider the ^{13}C nmr spectrum first.

^{13}C NMR Spectrum

We can make the following, general, assignments.

Chemical shift δ (ppm)	Integral	Assignment
166	1	C = O
136	1	Aromatic
134	1	Aromatic
132	1	Aromatic
131	1	Aromatic
130	2	Aromatic
129	2	Aromatic
128	2	Aromatic
126	2	Aromatic
65	1	C – O

This is a fascinating spectrum as there are twelve aromatic carbon atoms and confirms the presence of two linked but not fused aromatic rings. They must be separate from each other as fused aromatic systems such as naphthalene share carbon atoms and would only need ten aromatic carbon atoms whereas we have twelve to assign.

There are four pairs of chemically and magnetically equivalent carbon atoms and so, again, we have a potential structure such as:

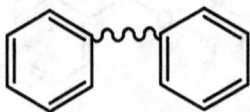

where the squiggle, ∿ , represents the remainder of the formula.

This makes sense since the ^{13}C nmr spectrum indicates that there are four pairs of chemically and magnetically equivalent aromatic carbon atoms (the four singlets of integral two) and four others which are all chemically and magnetically non-equivalent as indicated by the four singlets each of integral one. Two of these pairs will be on each of the aromatic rings which will each also contain two carbon atoms which are not chemically and magnetically equivalent to any other carbon atom.

the rest of the molecular formula comprises $C_2H_2O_2$ which must be an ester function meaning that the potential structure is:

We can now, confidently assign the signals as shown below where the pairs of chemically and magnetically equivalent are labelled a/a', b/b' and i/i', j/j' respectively. All the other carbon atoms are labelled alphabetically. This equivalence and non-equivalence becomes clearer if the page is rotated slightly.

C_f and C_g will be deshielded due to their proximity to the electronegative oxygen atoms and this deshielding might extend to the aromatic C_e and C_h

$C_a/C_{a'}$ and $C_i/C_{i'}$ will appear at similar positions and $C_b/C_{b'}$ and $C_j/C_{j'}$ will also appear at similar positions but it is not feasible to assign specific chemical shifts to these pairs since they are all quite close together although the carbon atoms closest to the electronegative oxygen atoms will be slightly deshielded. We can, therefore, tabulate, with notice taken of our inability to precisely assign all peaks as follows:

Chemical shift δ (ppm)	Integral	Assignment
166	1	C_f
136	1	C_e or C_h
134	1	C_e or C_k
132	1	C_c or C_k
131	1	C_c or C_k
130	2	$C_a/C_{a'}$ or $C_i/C_{i'}$
129	2	$C_a/C_{a'}$ or $C_i/C_{i'}$
128	2	$C_b/C_{b'}$ or $C_j/C_{j'}$
126	2	$C_b/C_{b'}$ or $C_j/C_{j'}$
65	1	C_g

Although we cannot precisely assign all chemical shifts we can be confident of the overall assignments simply because no other structure fits.

This means, of course, that we must also examine the ^1H nmr spectrum and, as we have a likely structure, it is sensible and, indeed, good practice to predict the spectrum before examining the observed spectrum. This is our next and final task.

^{1}H NMR Spectrum

If we consider the proposed structure again, with freshly labelled hydrogen atoms

we can make a number of following predictions. It is not too hard a task as the comparable aromatic hydrogen atoms will be split in the same way in both aromatic rings so we will focus on the left hand aromatic ring as the same principles will apply to the right hand aromatic ring.

- We can start with Ha:

The singlet due to H_a will be split into a doublet by H_b and this doublet will be split into a doublet of doublets by Hc

The singlet due to $H_{a'}$ will be split into a doublet by $H_{b'}$ and this doublet will be split into a doublet of doublets by Hc

- $H_{d'}$ will be split into a doublet by $H_{e'}$ and this doublet will be split into a doublet of doublets by H_e and, likewise, H_d will be split into a doublet by H_f and this doublet will be split into a doublet of doublets by H_e as shown below:

The singlet due to $H_{d'}$ will be split into a doublet by $H_{e'}$ and this doublet will be split into a doublet of doublets by H_f

The singlet due to H_d will be split into a doublet by He and this doublet will be split into a doublet of doublets by H_f

- With regard to H_b, $H_{b'}$, H_e and $H_{e'}$

  - H_b will be split into a triplet by H_a and H_c,
  - $H_{b'}$ will be split into a triplet by $H_{a'}$ and H_c,
  - H_e will be split into a triplet by H_d and H_f,
  - $H_{e'}$ will be split into a triplet by H_f and $H_{d'}$

as shown below:

The singlet due to H_b will be split into a triplet by H_a and H_c

The singlet due to $H_{e'}$ will be split into a triplet by $H_{d'}$ and H_f

The singlet due to $H_{b'}$ will be split into a triplet by $H_{a'}$ and H_c

The singlet due to H_e will be split into a triplet by H_d and H_f

58

Whilst not chemically and magnetically equivalent due to slightly different chemical and magnetic environments,

- H_a and H_d,
- H_b and $H_{e'}$,
- H_c and H_f,
- $H_{b'}$ and H_e,
- $H_{a'}$ and H_d

will produce similar chemical shifts and this means that the aromatic region will be extremely complex due to overlapping of the multiplets but it is important to note that Hc will produce a singlet, due to no hydrogen atoms on adjacent carbon atoms, of integral two.

If we examine the expanded ^1H nmr spectrum we will observe a number of peaks as shown below:

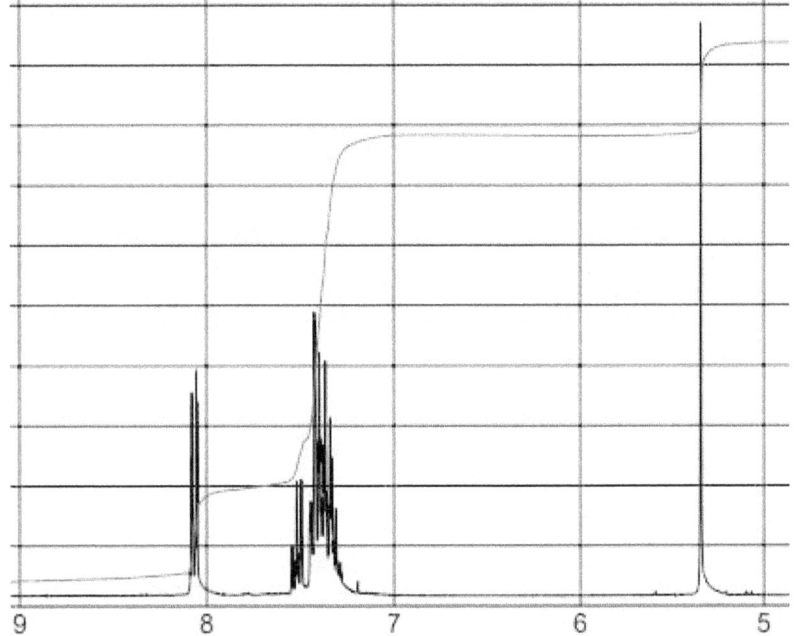

- The singlet at δ 5.35 ppm, of integral two, can be assigned to the two hydrogen atoms, – CH_2 –, labelled as H_c.
- The triplet, of integral two, at δ 8.10 ppm is due to either H_c or H_f.

The remainder of the aromatic hydrogen atoms, between δ 7 and δ 8 ppm, of integral eight, account for the complex multiplet which due to such enormous overlapping cannot be assigned but this concurs with the ^{13}C nmr, infra red and mass spectra.

We can now be confident that the assignments prove that the proposed structure is correct.

Conclusions

Structure:

Systematic name: Benzyl benzoate.

Chapter XI

Epinephrine

First isolated in 1895, ***epinephrine*** which is much popularly known as **adrenaline** and is found in the adrenal glands. It is known to play a significant role in the *fight or flight* response by increasing blood flow to muscles, heart output, pupil dilation and blood sugar levels.

It is used to treat many conditions including anaphylaxic shock, cardiac arrest and asthma and is administered intravenously, by direct injection into the muscles, by inhalation or by injection just below the skin.

Epinephrine production is stimulated by exercise and this phenomenon was first demonstrated by measuring the dilation of the pupils of a cat on a treadmill. The *fight or flight* or *emotional response* comprises two components, termed *autonomic* and *hormonal* and it is the hormonal response which releases epinephrine.

Epinephrine also plays a role in memory strengthening and has also been implicated in post traumatic stress disorder (PTSD).

Although epinephrine was first isolated by the American biochemist John Abel, in 1897, who coined the name, a purified extract was prepared by the chemist Jōkichi Takamine. This product was trademarked by the American pharmaceutical manufacturer, Parke, Davis & Co., under the name *adrenaline*.

Biologically, epinephrine is produced in a process starting with L – phenylalanine.

With a melting point of 211.5°C and a boiling point of 413°C, epinephrine is a colourless solid. With elemental composition C: 58.95%, H: 7.17%, N: 7.65%, O: 26.20% and formula mass 183.21 g mol^{-1} epinephrine has the **empirical** and **molecular** formulas, $C_9H_{13}NO_3$.

Infrared Spectrum

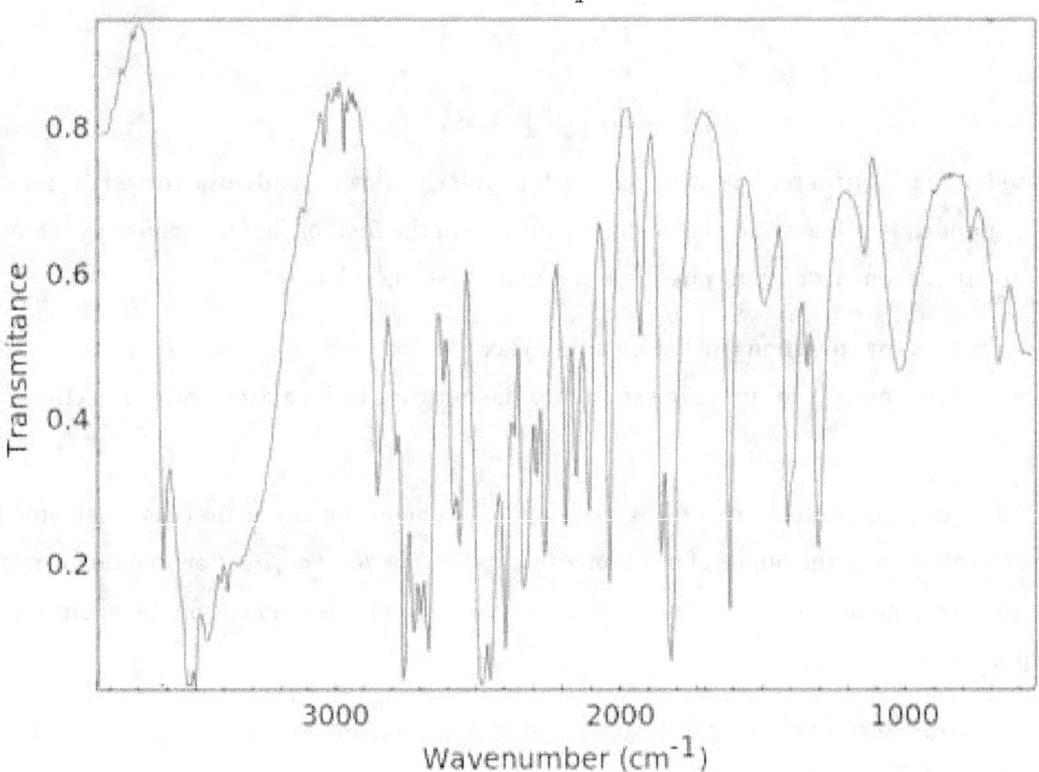

Observations

(√ / X)	Wavenumber range (cm⁻¹)	Wavenumber (cm⁻¹)	Assignment
√	3200 - 3700	3620	O – H
√	3200 - 3600	3520	N – H
√	3000 – 3300	3480, 3400	C – H (aromatic)
√	2500 – 3000	2880, 2780	C – H (aliphatic)
X	2200 – 2500		C ≡ N
X	1700 – 1800		C = O
X	1600 – 1700		C = C (aliphatic)
X	1585 – 1600		C – C (aromatic)
X	1450 – 1600		C – C (aromatic)
X	1000 – 1300		C – O
X	700 – 1000		C – X (X = Cl, Br or I)

Conclusions

This is a very complicated spectrum but it indicates:-

- That the molecule is aromatic and;

- It is significant that there are both O – H and N – H bonds.

- It cannot be either an ester or a carboxylic acid due to the absence of both C = O and C – O bonds.

- Since there are three oxygen atoms in the molecular formula it is reasonable to start from the basis that the molecule is a substituted phenol.

Mass Spectrum

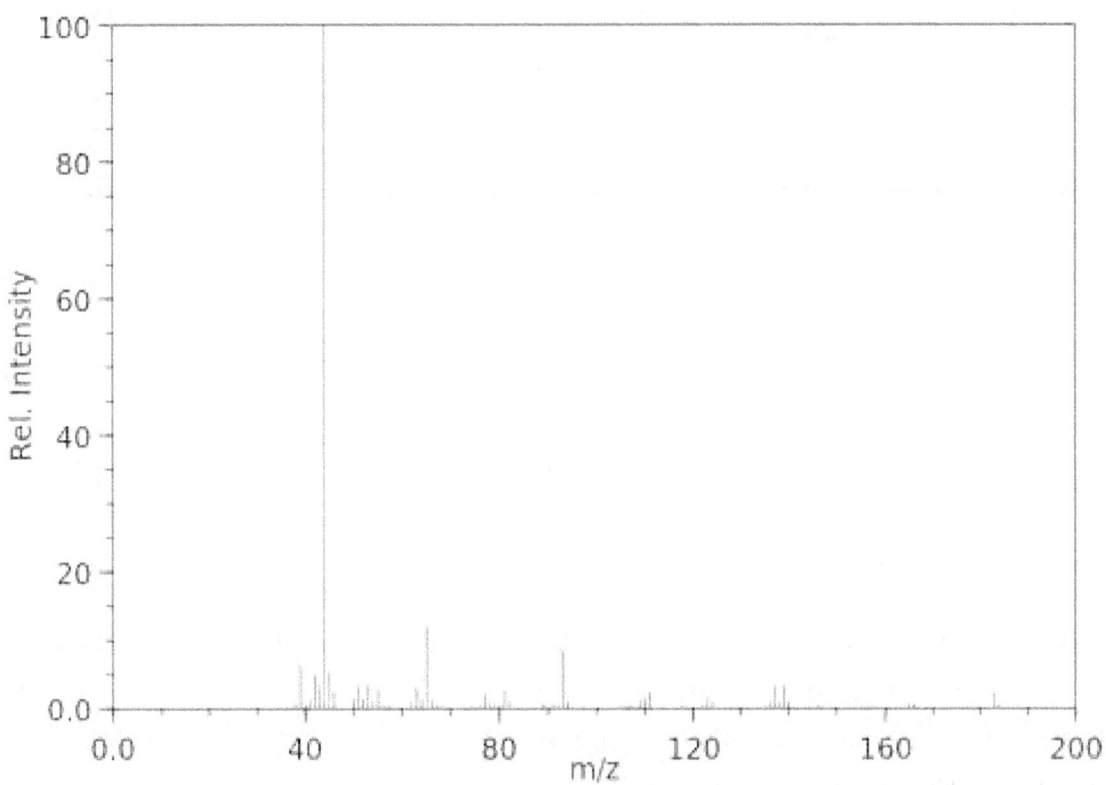

Observations

Charged fragments (m/z)	Assignment	Charged fragments (m/z)	Assignment
Molecular ion: 183	$[C_9H_{13}NO_3]^+$	Base peak: 44	$[C_2H_6N]^+$
139	$[C_7H_7O_3]^+$	93	$[C_7H_9]^+$
111	$[C_5H_7O_3]^+$	65	$[C_5H_5]^+$

Conclusions

- There are a number of peaks between m/z = 74 and 83 which implies that there is an aromatic ring in the compound. This concurs with the infrared spectrum which displays several aromatic C – H stretches.

- The base peak, m/z = 44, equates to a CH_3CHNH_2 or a CH_3CH_2NH functional group.

At this stage it is not possible to confidently assert more than this and we can learn far more from the ^1H and ^{13}C nmr spectra which is our next task.

NMR Spectra

The ^{1}H and ^{13}C nmr spectra are displayed below:

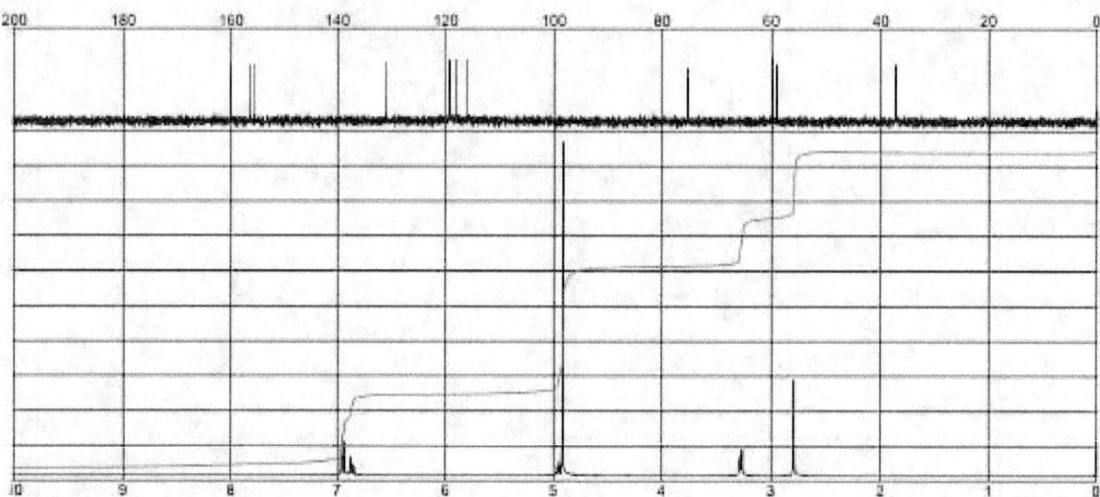

Both the ^{1}H and ^{13}C nmr spectra are clear with distinct, clear resonant signals. We have seen from previous chapters that the ^{13}C nmr spectrum is extremely useful in establishing the basic skeletal structure of the molecule. Since we can then complete the structure with the essential information provided by the ^{1}H nmr we will start by considering the ^{13}C nmr spectrum.

^{13}C NMR Spectrum

Chemical shift δ (ppm)	Integral	Assignment
157	1	Aromatic
156	1	Aromatic
132	1	Aromatic
119	1	Aromatic
118	1	Aromatic
116	1	Aromatic
75	1	C – O
59	1	C – N
37	1	C – N

The molecule clearly contains an aromatic ring as there are six aromatic carbon resonances and this molecule must have an aliphatic substituent containing the two nitrogen atom and the sole oxygen atom.

We can deduce a lot from the chemical shifts of the carbon atoms:

▪ If we examine the aromatic resonances in more detail we note that two, δ 157 ppm and δ 156 ppm, are strongly deshielded. This must be due to both having an electronegative element as substituent. There are three oxygen atoms and one nitrogen atom in the molecule and all three would deshield the aromatic carbon atoms. Since the two singlets are very close together, almost chemically and magnetically equivalent, this implies that the substituents are oxygen atoms which, in turn, implies that the ring has two hydroxyl, – OH, substituents.

- Since the other aromatic carbon atoms are not deshielded then this also implies that the remainder of the molecule forms a single aliphatic substituent and this must connect to the ring through bonding between an aromatic and an aliphatic carbon atoms – if it was bonded through either the nitrogen or oxygen atom then the aromatic carbon would also be deshielded.

- Since the aromatic ring contains three substituents, the two – OH groups and the aliphatic functional group this means that the aromatic ring will contain three hydrogen atoms on the ring itself and this means that the formula of the aromatic fragment and its two hydroxyl substituents is $C_6H_3(OH)_2$ – which means that the aliphatic substituent functional group will have the formula C_3H_8NO which we consider later.

Since there are three substituents on the ring, two O – OH and one aliphatic group denoted as R there are six possible general isomers as shown below:

where R is the aliphatic functional group of formula C_3H_8NO.

We will be able to determine the correct isomer from the ^1H nmr spectrum but our next task is to determine the structure of the aliphatic group, R.

Aliphatic functional group

There are three carbon atoms whose position we must determine.

- There is one C – O and two C – N bonds which cause the singlets at δ 75 ppm and δ 59 ppm and δ 37 ppm respectively. The latter two signals are significantly different but must be due to the C – N bonds as they are outside the C – O range and the, aliphatic, –C – OH peak has already been assigned (δ 75 ppm).

- As there is only one – C – OH peak then this must either be a substituent to a short chain or the terminal group on the short chain.

- There are a number of ways that we can draw a plausible structure for the aliphatic group as shown below:

Since there are two C – N bonds but and one C – O bond only isomers IV and V are plausible structures.

We will be able to distinguish between these two isomers through the multiplicities of the aliphatic hydrogen resonance signals and so our next task is to examine the ^1H nmr spectrum.

^{1}H NMR Spectrum

As before we can predict the spectrum of each of the possible isomers and we can break these down into two parts, prediction of the:

- Aromatic region of the ^{1}H nmr spectrum and then
- The aliphatic region.

We will start with the aromatic region.

Aromatic Region Analysis

There are five possible isomers of the aromatic ring but we can group them together.

In each of the following examples, the hydrogen atoms are labelled for that isomer only and the positions will be relabelled in each example.

Set I

- Isomers I and III both have three hydrogen atoms on immediately adjacent carbon atoms:

I   III

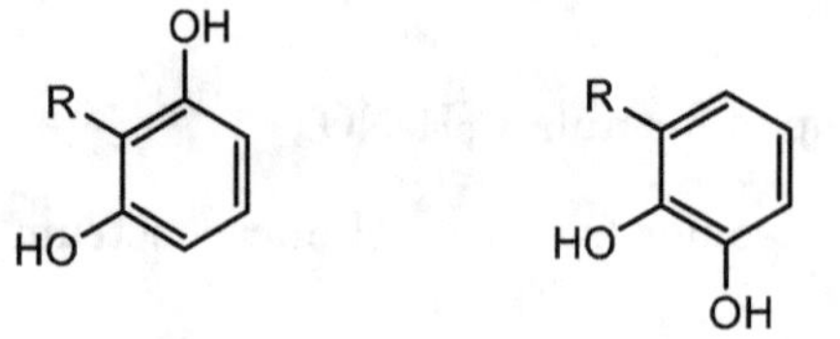

Both these isomers will produce three signals, each of integral one, and two will be doublets of doublets and one will be a triplet as shown below:

- **Isomer I:**

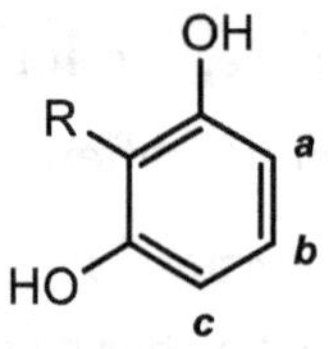

- H_a will be split into a doublet by Hb and into a doublet of doublets by H_c;
- Hb will split into a triplet by H_a and H_c;
- H_c will be split into a doublet by Hb and into a doublet by H_a.

- **Isomer III:**

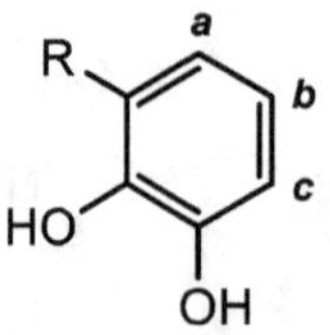

- H_a will be split into a doublet by H_b and into a doublet of doublets by H_c;
- H_b will split into a triplet by H_a and H_c;
- H_c will be split into a doublet by H_b and into a doublet by H_a.

Set II

Three isomers will produce the same pattern in the ^1H nmr spectrum

II   IV   VI

In each case, there will be one singlet and two doublets as demonstrated below:

Isomer II

- H_a will be split into a doublet by H_b;
- H_b will be split into a doublet by H_a;
- H_c will be a singlet as there are no hydrogen atoms on adjacent carbon atoms.

Isomer IV

- H_a will be a singlet as there are no hydrogen atoms on adjacent carbon atoms;
- H_b will be split into a doublet by H_c;
- H_c will be split into a doublet by H_b;

Isomer VI

- H_a will be split into a doublet by H_b;
- H_b will be split into a doublet by H_c;
- H_c will be a singlet as there are no hydrogen atoms on adjacent carbon atoms;

Set III

There is only one isomer, V, in this set.

This molecule will produce three singlets all of integral one since there are no hydrogen atoms on any adjacent carbon atoms.

If we now examine the aromatic region of the ¹H nmr spectrum we can establish that there are multiplets due to the three hydrogen atoms in the ring but there is also another signal whose identity is to be determined:

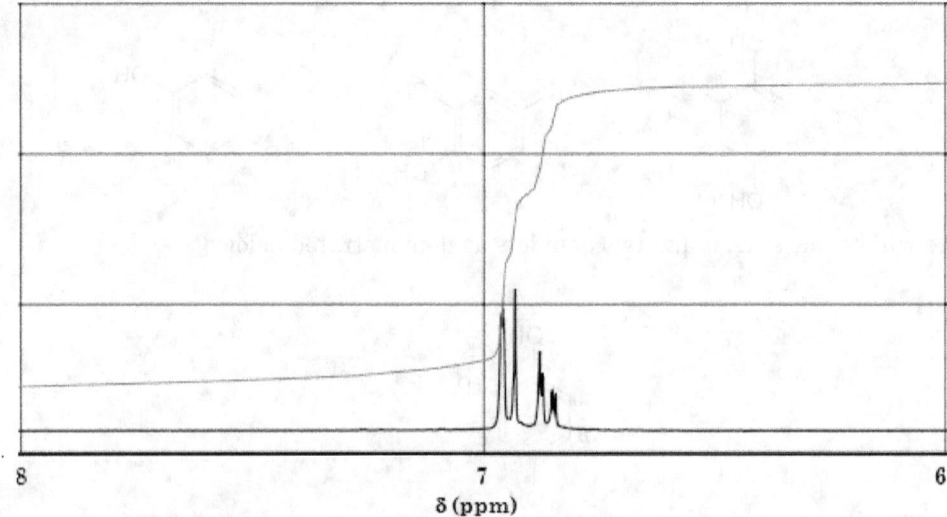

There is a singlet of integral one at δ 6.90 ppm and two doublets, each of integral one, at δ 6.88 ppm and δ 6.84 ppm. The triplet at δ 6.95 ppm is yet to be assigned.

The existence of a singlet and two doublets immediately reduces the number of possible candidate isomers to candidates II, IV and VI.

- In the case of **Isomer II**, all three hydrogen atoms are bonded to carbon atoms which are all adjacent to carbon atoms with hydroxyl functional groups so the peaks will all be significantly deshielded. This deshielding of all three hydrogen atoms is not observed.

- With **Isomer IV**, two hydrogen atoms are on carbon atoms with hydroxyl functional groups on adjacent carbon atoms and one, adjacent to the carbon atom with the aliphatic group, R, attached and asterisked is remote from the two – OH groups

This means that the singlet is deshielded as is one of the doublets whilst the asterisked hydrogen atom is not deshielded. This is what we observe.

- **Isomer III** will have a deshielded singlet as it is between two carbon atoms with hydroxyl functional groups. This is not observed.

This means that the only plausible structure is **Isomer IV** but to ensure that this structure is correct we also need to examine the rest of the ¹H nmr spectrum which will be due to the aliphatic functional group substituent.

Aliphatic Region Analysis

Once again, it is wise to predict the spectrum of our candidate molecule before examining the observed spectrum.

We have two possible structures which are labelled A and B below:

A   B

- Both molecules are chiral since the asterisked carbon atom has four different substituents. The hydrogen atom on the asterisked carbon atom will be of integral one and will be a singlet as there is only one adjacent carbon atom and this aromatic carbon atom does not have any hydrogen atoms bonded to it. That there are hydrogen atoms in the – OH and – NH is irrelevant since the coupling does not extent across these atoms but, since they are electronegative they will deshield the hydrogen atom on the chiral carbon atom.

- With regard to the remaining hydrogen atoms in the functional group, there may or may not be signals due to the – OH and – NH hydrogen atoms as they are labile but if they do appear then they will both produce, in both isomers, singlets of integral one. These singlets can appear anywhere in the range δ 0 – 12 ppm.

- Examining the alkyl hydrogen atoms,
    - **Isomer A** contains a – CH_2 – CH_3 group

        - H_a will produce a quartet of integral two due to the presence of the adjacent methyl group (H_b). These hydrogen atoms will be deshielded due to the presence of the adjacent, electronegative nitrogen atom and so will appear downfield, i.e. of higher chemical shift, than for a fully aliphatic – CH_2 –. This is likely to appear somewhere in the region δ 3 – 4.5 ppm.

        - H_b will generate a signal of integral three and it will be a triplet due to the two, H_a, hydrogen atoms. This will not be deshielded since it is too far from the nitrogen atom and so will appear in the δ 0.5 – 2 ppm region.

        - H* will be a singlet of integral one.

Isomer B

- Contains a – CH_2 – group (H_c) which will produce a doublet of integral two, the doublet being caused by splitting by the hydrogen atom bonded to the chiral carbon atom. This will be deshielded due to the adjacent – NH – group.
- Contains a methyl, – CH_3, group (H_d) and the signal due to this group will be deshielded due to the proximity of the – NH – group meaning that it should appear somewhere between δ 3 and δ 4.5 ppm. It will be a singlet since there are no adjacent carbon atoms.
- H* will, again, be a singlet of integral one.

We must now re-examine the actual ^1H nmr spectrum

We can immediately identify the correct isomer by observing the singlet of integral three at δ 4.9 ppm, the singlet of integral two at δ 2.8 ppm and a doublet of integral one at δ 3.3 ppm. Only **Isomer B** meets this observation and so the molecule must have the structure:

Conclusions

Structure:

Systematic name: (R)-4-(1-Hydroxy-2-(methylamino)ethyl)benzene-1,2-diol

It has already been noted that the molecule is chiral. It is administered as a racemic mixture and the two enantiomers have optical rotation of +/- 53º.

Chapter XII

Warfarin

Originally produced as a rat poison in 1948, ***warfarin*** is used as a blood thinner which is technically known to be an anticoagulant. It is also used to treat deep vein thrombosis, pulmonary embolism as well as being prescribed to prevent strokes in people with heart disease or a history of heart attacks. Usually administered orally, in tablet form, it can also be administered intravenously.

As with all pharmaceuticals, its side effects include intestinal bleeding and it is not prescribed for pregnant women. It was recommended as a blood thinner in 1954 and its reputation was bolstered the following year when, the then, President, Dwight D. Eisenhower was treated with it following a heart attack.

With a melting point of approximately 155°C and a boiling point of 515°C, this compound is, at room temperature, a colourless, crystalline solid.

Warfarin has the elemental composition: C: 73.95, H: 5.24%, O:20.76% and it has the **formula mass** (M_r) of 305.50 g mol^{-1}.

This means that the **empirical** and **molecular** formulas are both $C_{19}H_{16}O_4$.

Infrared Spectrum

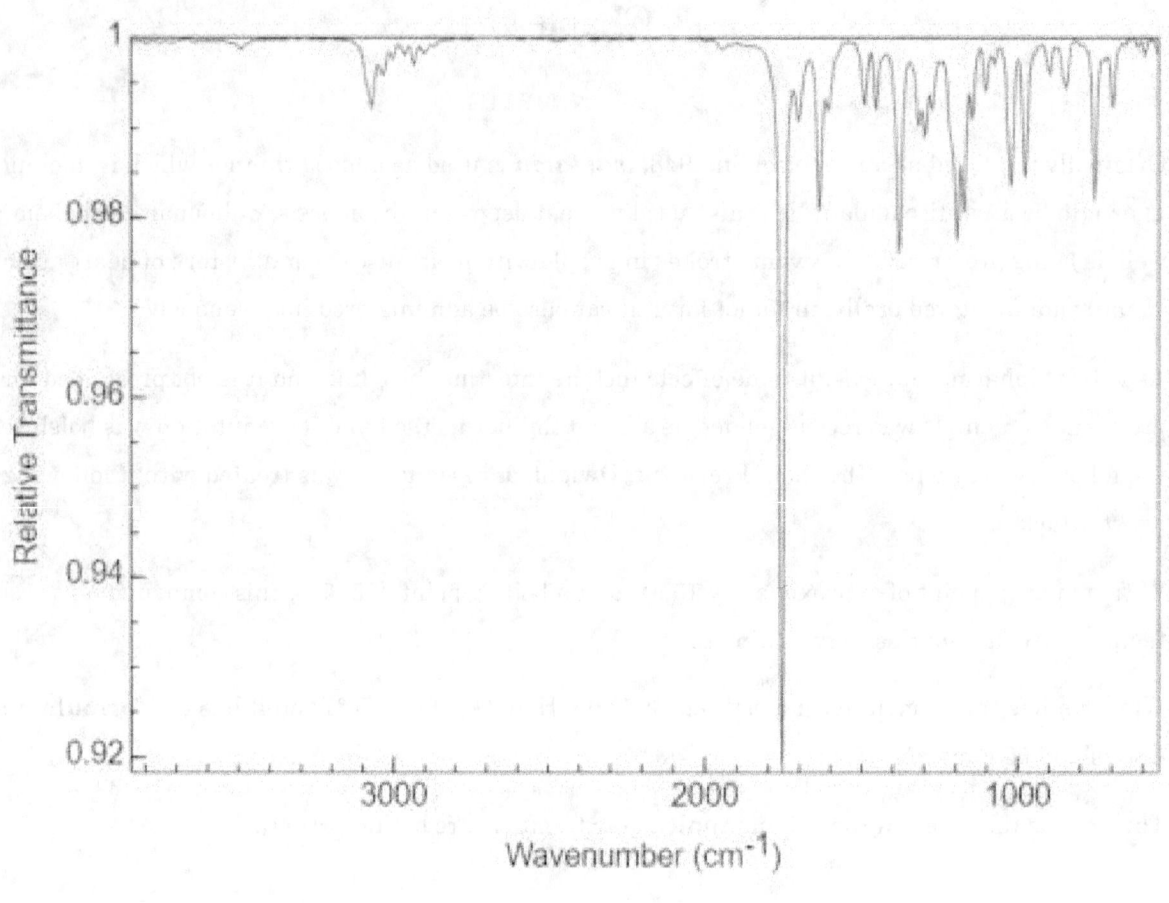

Observations

(√ / X)	Wavenumber range (cm⁻¹)	Wavenumber (cm⁻¹)	Assignment
√	3200 - 3700	3500	O – H
X	3200 - 3600		N – H
√	3000 – 3300	3070, 3010	C – H (aromatic)
√	2500 – 3000	2990, 2920, 2910, 2890	C – H (aliphatic)
X	2200 – 2500		C ≡ N
√	1700 – 1800	1760	C = O
√	1600 – 1700	1620	C = C (aliphatic)
X	1585 – 1600		C – C (aromatic)
√	1450 – 1600	1485	C – C (aromatic)
√	1000 – 1300	1200	C – O
X	700 – 1000		C – X (X = Cl, Br or I)

Conclusions

This compound contains:-

- At least one hydroxyl group;

- at least one aromatic ring, aliphatic functional groups and;

- Eeither an ester or a carboxylic acid functional group or both.

- There is also at least one C = C bond.

Mass Spectrum

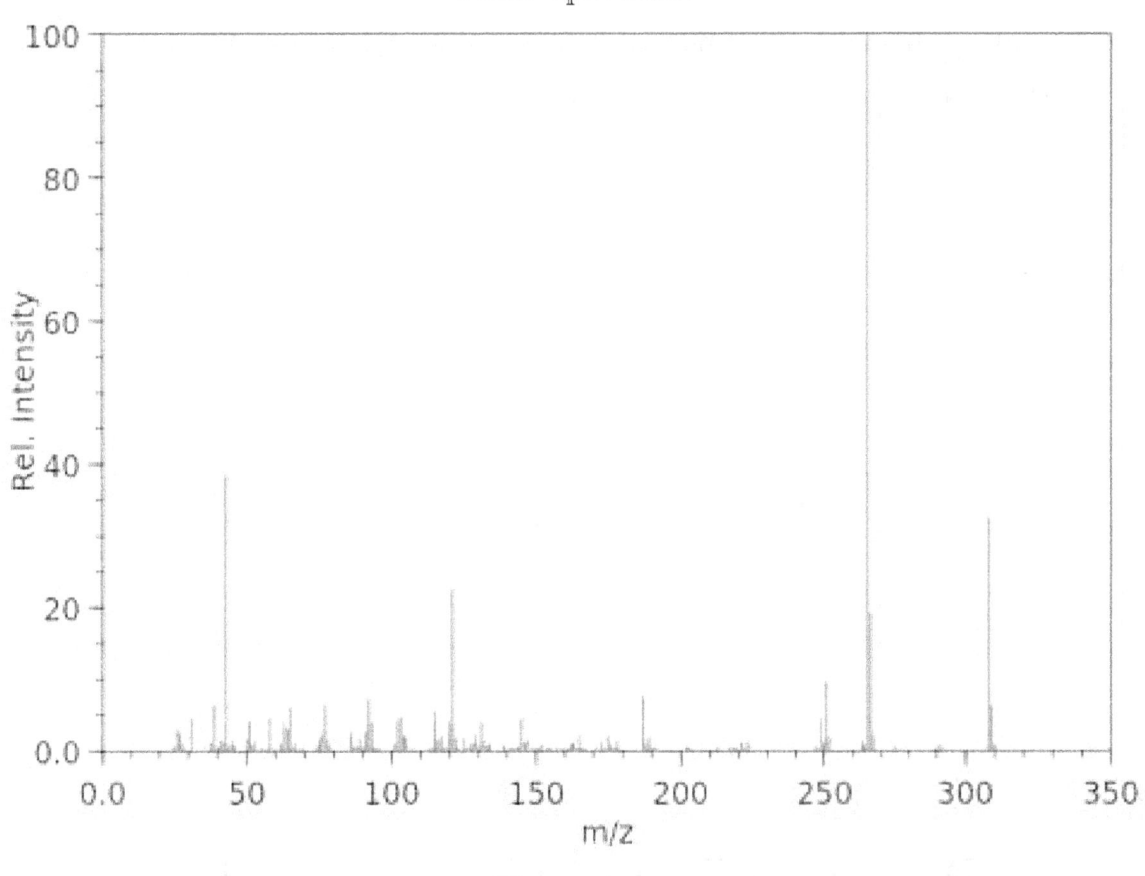

Observations

Charged fragments (m/z)	Assignment	Charged fragments (m/z)	Assignment
Molecular ion: 308	$[C_{19}H_{16}O_4]^+$	Base peak: 265	$[C_{16}H_9O_4]^+$
251	$[C_{15}H_7O_4]^+$	77	$[C_6H_5]^+$
223	$[C_{15}H_7O_2]^+$	65	$[C_5H_5]^+$
187	$[C_{15}H_7]^+$	58	$[C_4H_{10}]^+$
103	$[C_6H_5 - CH_2 - C]^+$	43	$[C_3H_7]^+$
92	$[C_6H_5 - CH_2 + H]^+$	39	$[C_3H_3]^+$

Conclusions

▣ The most significant peak is m/z 77 which is always due to a benzene ring with one substituent.

▣ It is not possible to determine the source of many of the fragments as they may be due to fragmentation of the ring(s) or fragmentation of the functional group(s) or both.

▣ The molecular formula, $C_{19}H_{16}O_4$, implies the existence of more than one aromatic ring as there are insufficient hydrogen atoms for the rest of the molecule to be aliphatic.

NMR Spectra

Since the ^{13}C NMR spectrum contains a number of clear signals in contrast to the messy looking ^{1}H nmr spectrum it is clearly sensible to start by examining this first.

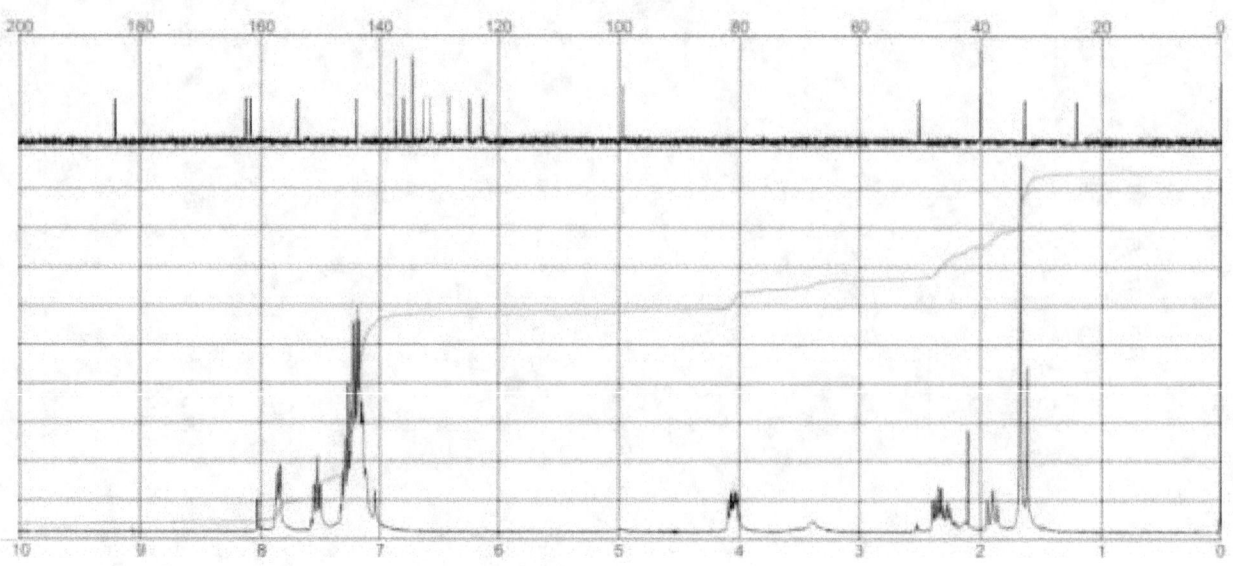

^{13}C NMR Spectrum

Observations

Chemical shift δ (ppm)	Integral	Assignment
185	1	C = O
157	1	C = C or C – C aromatic
156	1	C = C or C – C aromatic
149	1	C = C or C – C aromatic
145	1	C = C or C – C aromatic
137	2	C = C or C – C aromatic
136	1	C = C or C – C aromatic
134	2	C = C or C – C aromatic
132	1	C = C or C – C aromatic
131	1	C = C or C – C aromatic
129	1	C = C or C – C aromatic
125	1	C = C or C – C aromatic
122	1	C = C or C – C aromatic
99	1	C – O
50	1	C – C
34	1	C – C
25	1	C – C

There are fourteen signals that can be assigned to aromatic rings and alkenes. Since there are fewer hydrogen atoms than carbon atoms there are likely to be two benzene rings in the molecule which could account for twelve of the carbon atoms leaving two carbon atoms to form an alkene.

Whilst not assigning any stereochemistry to the molecule yet this leaves us to account for five carbon atoms, four oxygen atoms and the remaining five hydrogen atoms. From $C_5H_5O_4$, we have to find an ester and a hydroxylic acid group or a carboxylic acid group whose presence has been identified from the *infrared spectrum*.

Since the mass spectrum contains a peak at m/z = 77 the molecule must contain a benzene ring with a single functional group and so the basis of the molecule can be considered to be:

This now allows to concentrate on the $C_{13}H_{11}O_4$ group which must contain another benzene ring but we need to use up some more of the carbon atoms and also a C – O bond. The carboxylic acid or ester grouping must form a terminal group on an aliphatic chain. There is still an excess of carbon atoms compared to hydrogen atoms and this implies that there is a third ring which is fused to the second benzene ring.

This now gives us a possible base molecule:

which cannot be completely correct as the $C_3H_4O_4$ link cannot be constructed in a linear way to match the spectra.

We can examine the ^{13}C nmr spectrum again but this time using the chemical shift of the carbon atoms in benzene itself (δ 128.5 ppm) and can predict that there will be sixteen carbon signals close to the base, benzene, signal. This does not occur and there are only nine in the region δ 125 – 137 ppm which are not shifted upfield or downfield far from the base, chemical shift of benzene.

There are two singlets of integral two, δ 137 and δ 134 ppm, and a singlet at δ 129 ppm, which are standard for a monosubstituted benzene ring. This leaves us with four carbon atoms which are largely neither shielded nor deshielded and so we can, tentatively, make the following assignments of these peaks but at this stage cannot precise assignments. Nevertheless, the following is plausible:

C_e / C_h and C_f / C_g can account for the two singlets of integral two whilst the singlet at δ 129 ppm can be assigned to C_i *if* the ring is joined to the molecule by another carbon atom.

This leaves us with five aromatic carbon atoms to be assigned. Four carbon atoms are significantly deshielded at δ 157, 156, 149 and 145 ppm and so these must have electronegative i.e. oxygen atoms attached or bonded.

One plausible structure is where an oxygen atom is in the unlabelled ring. This oxygen atom could be in one of three places as highlighted below:

If any of these are correct, the oxygen – containing ring will no longer be aromatic but will still account for another three signals in the ^{13}C nmr spectrum.

- **Isomer I** cannot be correct as the alkene para to the oxygen atom will produce two chemically and magnetically equivalent alkene carbon signals and this is not observed;
- **Isomer II** cannot be correct since it will produce two C – O signals and this is not observed which leaves us with:
- Isomer **III** whose structure, for convenience, is repeated below:

The two alkene carbon atoms will produce resonance signals close to each other if the squiggly arrow represents a linkage starting and ending with a carbon atom. There is however the possibility that the left hand carbon atom of the C = C bond is bonded to something other than a hydrogen atom. The only possible substituent would be a hydroxyl group which will deshield the bonded alkene carbon atom.

Extending the labelling to this structure and adding in the hydroxyl functional group, we can now portray the, partial, structure as

The bonding of a hydroxyl group to C_j accounts for one of the deshielded carbon signals (δ 157 ppm or δ 156 ppm) and this also means that, assuming that the bond between C_k and the right hand benzene ring begins with a carbon atom on the squiggly line, this structure also accounts for one of the lesser deshielded resonant signals (δ 149 ppm or δ 145 ppm). This is a little way away from the benzene carbon signal (δ 128.5 ppm) because it will be partially deshielded by its proximity to the hydroxyl group on C_j.

This leaves us with two signals in the δ 160 – δ 120 ppm region to assign: one of the pair δ 157 ppm / δ 156 ppm and one of the pair δ 149 ppm / δ 145 ppm. These can be assigned to the previously unlabelled carbon atoms in the fused ring as shown below where the previous unlabelled carbon atoms are now C_m and C_n:

This leaves us with the following cumulative formula to assign: C_4H_6O to assign.

To assign the remaining atoms simply requires us to work out the composition of the bonding represented by the squiggle.

We know that there is either a carboxylic acid or an ester functional group and these can only appear as a terminal functional group.

There are too many hydrogen atoms for a carboxylic acid functional group and so this means that the only possible structure is as shown below:

and we can now assign the remaining signals as follows:

C_r, C_p, C_o at δ 50, 34, 25 ppm respectively and the carbonyl, C_q, at δ 185ppm.

This fully accounts for the structure with the following assignments:

Chemical shift δ (ppm)	Integral	Assignment
185	1	C_q
157	1	C_l or C_i
156	1	C_h
149	1	Ck or
145	1	Ck or
137	2	C_e /C_i or C_f / C_h
136	1	C_m
134	2	C_e /C_i or C_f / C_h
132	1	C_a, C_b, C_c or C_d
131	1	C_a, C_b, C_c or C_d
129	1	C_g
125	1	C_a, C_b, C_c or C_d
122	1	C_a, C_b, C_c or C_d
99	1	C_n
50	1	C_r
34	1	C_p
25	1	C_o

We now need to confirm this structure by examining the ^1H nmr spectrum.

^{1}H NMR Spectrum

The ^{1}H nmr spectrum looks complicated and so it is probably best to use the proposed structure to predict the spectrum and then compare the prediction with the observed spectrum.

Starting with new alphabetical labelling, we have the following structure:

We can make the following overall predictions with considering any coupling or splitting:

- There will be nine peaks in the aromatic region (δ 6 – 8 ppm) due to the hydrogen atoms on the left hand ring, H_a, H_b, H_c, H_d and on the right hand ring, H_e, H_f, H_g, H_h and H_i:

    - H_a and H_c are mutually chemically and magnetically equivalent and will produce a multiplet of integral two.
    - H_b and H_d will produce multiplets both of integral one which will probably overlap.
    - H_e and H_i are mutually chemically and magnetically equivalent so will produce a multiplet of integral two.
    - H_f and H_h are also mutually chemically and magnetically equivalent so will also produce a multiplet of integral two.
    - H_g will produce a multiplet of integral one.

- There may or may not be a peak assignable to H_k of integral one – the hydrogen atoms are labile and are often not observable. If one does appear it will appear downfield, above, δ 10 ppm.

- H_j will produce a triplet of integral one, which must be due to splitting by the two H_l hydrogen atoms, in the region δ 0.5 – 2 ppm.

- H_l will produce a doublet, of integral two, due to splitting by H_j and this will appear in the region δ 2 – 3 ppm due to its proximity to the carbonyl group.

- Similarly, H_m will produce a singlet of integral three also in the δ 2 – 3 ppm region.
    The signal will be a singlet due to the absence of any hydrogen atoms on the adjacent carbon atom and in the range δ 2 – 3 ppm due to its proximity to the carbonyl group.

    This all seems to work and so we can summarise everything as shown below.

To summarise:

Chemical shift δ (ppm)	Integral	Multiplicity	Assignment
6 – 8	2	Doublet of doublets of doublets	H_a / H_c
6 – 8	1	Doublet of doublets of doublets	H_b
6 – 8	1	Doublet of doublets of doublets	H_d
6 – 8	2	Doublet of doublets of doublets	H_e / H_i
6 – 8	2	Doublet of doublets of doublets of doublets	H_f / H_h
6 – 8	1	Doublet of doublets of doublets of doublets	H_g
10 – 12	1	Singlet	H_k
0.5 – 2	1	Triplet	H_j
2 – 3	2	Doublet	H_l
2 – 3	3	Singlet	H_m

All of these are observed in the ^1H nmr spectrum.

Conclusions

Structure:

Systematic name: (RS)-4-Hydroxy-3-(3-oxo-1-phenylbutyl)-2H-chromen-2-one

Optical activity

This molecule exists as two enantiomers due to the chiral carbon atom identified by * – there is a hydrogen atom on the carbon atom and so this carbon atom, C_o, has four different substituents.

The two enantiomers are shown below:

R – (+) enantiomer   **S – (–) enantiomer**

Angle of rotation (°)	+ 149	- 149

Both enantiomers are effective but the S enantiomer is four times more potent than the R enantiomer and so the substance is administered as a racemic mixture i.e. in equal quantities.

Chapter XIII

Ephedrine

Ephedrine is a medication and stimulant which is often used to prevent low blood pressure and as treatments for asthma, narcolepsy, and obesity. It is also a popular, legal, supplement taken by bodybuilders seeking to cut body fat before a competition.

Common side effects include motion sickness, insomnia, anxiety, headaches, migraines, hallucinations and high blood pressure whilst other serious side effects include strokes and heart attacks.

First isolated from the ephedra, genus of gymnosperm shrubs. and synthesised in 1885, ephedrine came in to commercial use in 1926 it is now available as a cheap, generic medication.

It is commonly used in Chinese medicines as *ephedra sinica* and is sometimes referred to as *the oldest medicine*.

Structurally, it is similar to amphetamines and is a precursor to methamphetamine which is an illegal narcotic.

Ephedrine has a melting point of 36°C and a boiling point of 225°C and at room temperature it is a colourless, crystalline solid.

It has the elemental composition: C: 72.62%, H: 9.17%, N: 8.48%, O: 9.68% and its formula mass (M_r) is 165.24 g mol^{-1}. This means that the **empirical** and **molecular** formulas are both $C_{10}H_{15}NO$.

The compound comprises two optically active isomers, enantiomers and it is prescribed as a racemic mixture of the two optically active isomers.

Infrared Spectrum

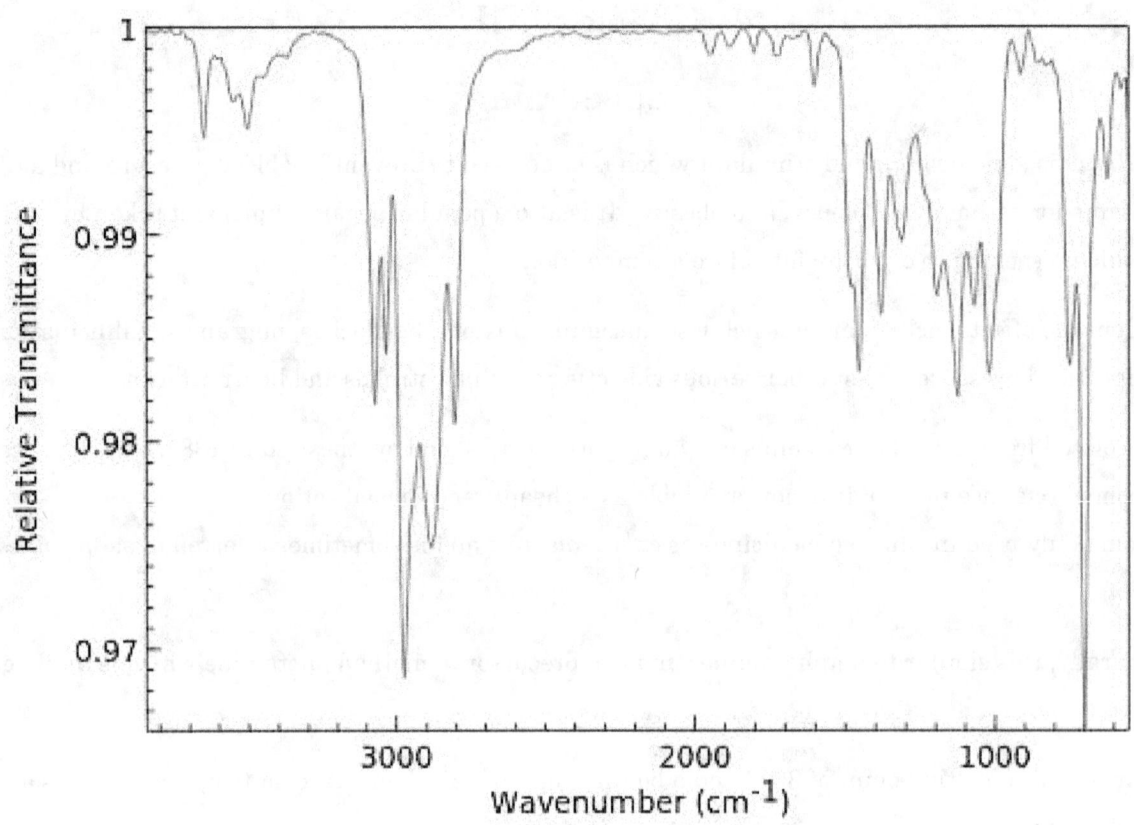

Observations

(√ / X)	Wavenumber range (cm⁻¹)	Wavenumber (cm⁻¹)	Assignment
√	3200 - 3700	3650	O – H
√	3200 - 3600	3500	N – H
√	3000 – 3300	3090, 3040	C – H (aromatic)
√	2500 – 3000	2980, 2885, 2800	C – H (aliphatic)
X	2200 – 2500		C ≡ N
X	1700 – 1800		C = O
X	1600 – 1700		C = C (aliphatic)
X	1585 – 1600		C – C (aromatic)
√	1450 – 1600	1450	C – C (aromatic)
√	1000 – 1300	1370	C – O
?	700 – 1000	Unassignable	C – X (X = Cl, Br or I)

Conclusions

▪ This compound contains an O – H bond, a N – H bond and is both aromatic and aliphatic;

▪ The molecule cannot be a carboxylic acid or an ester as there is no C = O bond and this suggests that the molecule might be a substituted phenol with or without an amino substituent elsewhere on the ring.

It is worth also noting that the strongest peak at 700 cm⁻¹ cannot be assigned.

Mass Spectrum

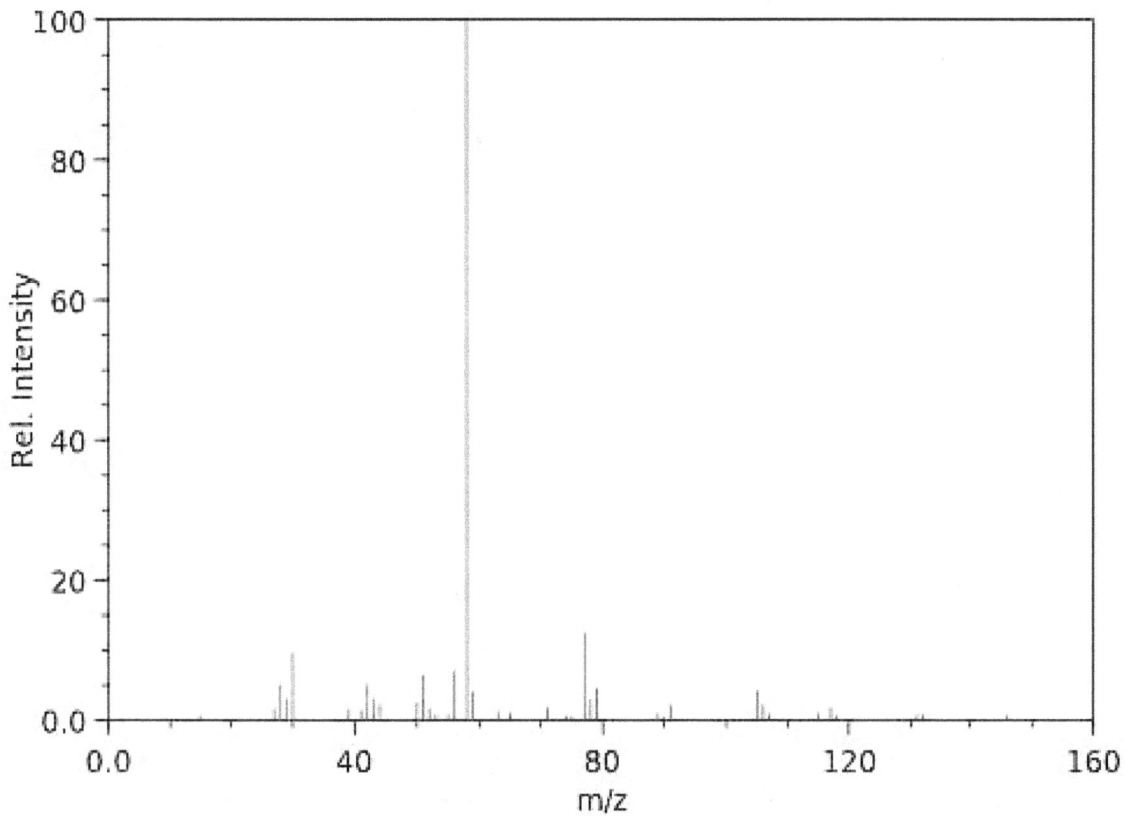

Observations

Charged fragments (m/z)	Assignment	Charged fragments (m/z)	Assignment
Molecular ion: 146	$[C_{10}H_{15}NO]^+$	Base peak: 58	$[C_3H_7CH_3]^+$
132	$[C_6H_5CH_2CH_2CH_3C]^+$	56	$[C_4H_8]^+$
117	$[C_6H_5CH_2CH_2C]^+$	51	$[C_4H_3]^+$
105	$[C_6H_5CH_2CH_2]^+$	43	$[C_3H_7]^+$
91	$[C_6H_5CH_2]^+$	39	$[C_3H_3]^+$
77	$[C_6H_5]^+$	15	$[CH_3]^+$

Conclusions

▣ There are a number of ways we can assign different fragments but the most important peak in the spectrum is that at m/z = 77 which is only ever due to $[C_6H_5]^+$ and indicates that this molecule has a benzene ring with only one substituted functional group.

▣ This functional group must have the overall formula $C_4H_{10}NO$ and a formula mass of 88 g mol^{-1}.

Since the molecule is a mono-substituted benzene ring it cannot be a phenol and so we have this as a starting place

$C_4H_{10}NO$

and it is now our task to determine the structure of the substituent from the ^1H and ^{13}C NMR spectra.

NMR Spectra

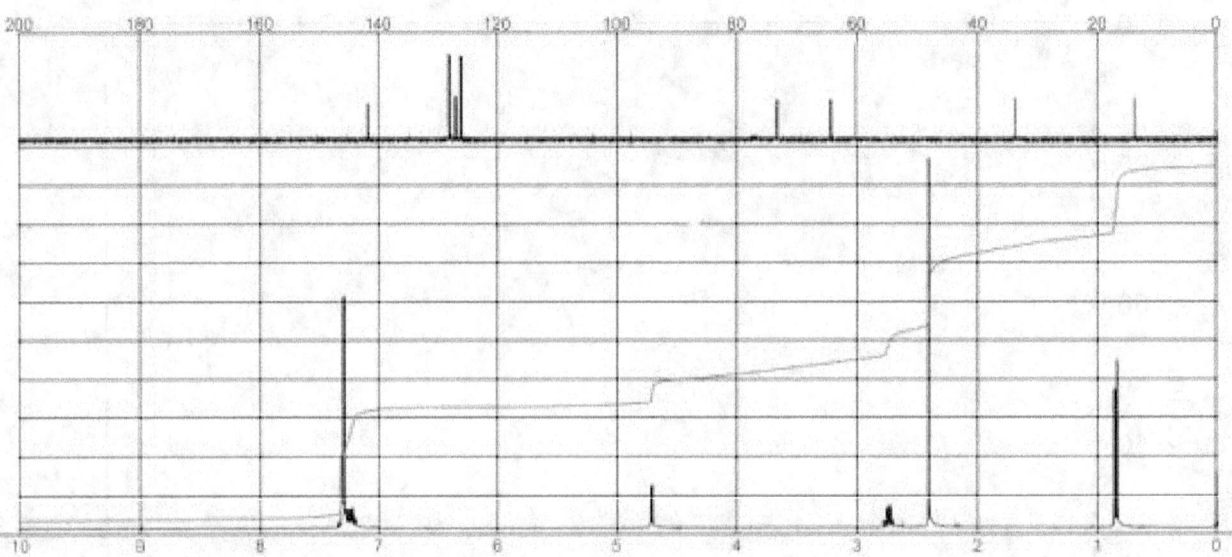

We will consider the ^{13}C nmr spectrum first making assignments from the data table.

^{13}C NMR Spectrum

Chemical shift δ (ppm)	Integral	Assignment
143	1	Aromatic
128	2	Aromatic
127	1	Aromatic
126	2	Aromatic
75	1	C – O
65	1	C – N
35	1	C – N
14	1	C – C

Observations and Conclusions

▪ The spectrum confirms the presence of a mono-substituted aromatic ring.

▪ There must be a C – C bond between the substituent group and the benzene ring;

▪ There are two C – N bonds and one N – H bond so the terminal group must be C – N(H) – CH3;

This means that we can draw a template as below where the squiggle represents the as yet undetermined remaining atoms:

The squiggle represents the remaining atoms: CH6O and the only way we can construct a full structure is to have a branched structure meaning that the full structure must be as follows:

If we add in the hydrogen atoms on the aliphatic functional group only we can predict the following:

This molecule has two chiral centres which are marked with an * and so, given that there are 2^n enantiomers for every **n (where n is an integer)** chiral centres, this means that there are four enantiomers of the molecule as, in this case, n = 2. Remember that for a carbon atom to be chiral it must simply have four different substituents. It doesn't matter how far along a substituent the difference is.

If we examine the aromatic region we observe that there are two singlets of integral two, one singlet of integral one and one deshielded singlet of integral one.

If we consider the molecule again but with all the carbon atoms labelled alphabetically we can more precisely assign the peaks:

We can see that there are two pairs of chemically and magnetically equivalent carbon atoms labelled as f / f' and g / g' and so the final assignment, based on the ^{13}C nmr spectrum is as follows:

Chemical shift δ (ppm)	Integral	Assignment
143	1	C_e
128	2	f / f' or g /g'
127	1	C_h
126	2	f / f' or g /g'
75	1	C_d
65	1	C_b
35	1	C_a
14	1	C_c

We must now confirm this structure using the ^1H nmr spectrum.

^{1}H NMR Spectrum

We can make a number of predictions from the proposed structure. Drawing the structure again but with newly labelled hydrogen atoms we have two ways of portraying the structure – with and without the bonded hydrogens. They are both displayed below and, for clarity, we will use the structure without all the hydrogen atoms displayed:

We can make some initial predictions:

- As shown, there are two pairs of chemically and mutually equivalent aromatic hydrogen atoms (H_g / $H_{g'}$ and H_h / $H_{h'}$). Both H_g and $H_{g'}$ will produce doublets of doublets due to:
  - H_g being split into a doublet by H_h and then into a doublet of doublets by H_i.
  - $H_{g'}$ being split into a doublet by $H_{h'}$ and then into a doublet of doublets by $H_{i'}$.

  These doublets of doublets will appear at the same chemical shift.
- H_h and $H_{h'}$ will produce triplets of integral one due to H_h being split by H_g and H_i and $H_{h'}$ being split by $H_{g'}$ and $H_{h'}$ respectively and they will appear at the same chemical shift so we will, again, observe a signal of integral two;

- H_i will produce a doublet of doublets of integral one due to splitting by H_h / $H_{h'}$ (producing a doublet) and then a doublet of doublets of doublets by H_g / $H_{g'}$;
- H_e and H_b may not appear in the spectrum due to the labile nature of the protons;
- H_a will appear as a singlet of integral three;
- H_f will appear as a doublet of integral three due to splitting by H_c;
- H_d will be a doublet of integral one due to splitting by H_c whilst;
- H_c will produce a signal of integral one which is a doublet of triplets due to splitting by H_f (producing a triplet) and then by H_d, producing a doublet of triplets.

The aromatic region ^{1}H nmr spectrum is a mess as so many peaks are overlapping but the significant information we can obtain from it is that there are two groups of overlapping multiplets of integral 3:2 and we can assign them to H_h / $H_{h'}$ / Hi (integral of three) and H_g / $H_{g'}$ (integral of two)

If we examine the rest of the spectrum we note that there is:

- A doublet of integral one at δ 4.75 ppm which can be assigned to H_d as there is only one hydrogen atom attached and it is also bonded to – OH.

- H_c is bonded to a carbon atom with three hydrogen atoms (H_f) and a carbon atom with one hydrogen atom attached (H_d) and so will produce a doublet of quartets due to the signal being split into a quartet by H_f (n+1 rule) and then this will be split into a doublet of quartets by H_d. This appears at δ 2.75 ppm and at low resolution this would simply appear as an octet.

- The singlet at δ 2.45 ppm, of integral three, is due to a methyl group and may be assigned to H_a.

- There is a doublet of integral three at δ 0.88 ppm which can be assigned to H_f.

All of these assignments are consistent with the proposed structure and we can summarise the assignments in the ^{1}H nmr spectrum as follows:

Chemical shift δ (ppm)	Integral	Multiplicity	Assignment
7.35	3	Triplet	H_h /$H_{h'}$ / H_i
7.25	2	Complex multiplet	H_g / $H_{g'}$
4.75	1	Doublet	H_d
2.75	1	Doublet of quartets	H_c
2.45	3	Singlet	H_a
0.88	3	Singlet	H_f

There are two signals missing and these will be due to the labile H_b and H_e atoms.

Conclusions

Structure:

Systematic name: (1R,2S)-(–)-Ephedrine

Optical activity:

As mentioned before, this molecule has two chiral carbon atoms (H_c and H_d) and the angle of rotation is -34^0. It is not possible to predict this and it can only be determined by direct measurement.

The designations R and S relate to the stereochemistry using the Cahn-Ingold-Prelog rules which have been discussed in previous volumes in this series.

Chapter XIV

Nicotinamide

Nicotinamide is a form of the essential nutrient, vitamin B3.

Natural food sources include red meat, poultry, tuna and salmon and, in lesser quantities, in nuts, legumes and seeds. Nicotinamide also used as a dietary supplement for treating the disease *pellagra* which is caused by niacin deficiency and has symptoms including skin and mouth lesions, anaemia and migraines. In many countries, niacin is added to wheat flour to reduce instances of pellagra.

Discovered in 1935, nicotinamide is available over the counter and is also used, as a cream, as an acne treatment. With minimal side effects, it is also used a preventative treatment for those at high risk of melanoma.

Nicotinamide has a melting point of 130°C and a boiling point of 334°C and it is a colourless, crystalline solid.

With an elemental composition of C: 58.95%, H: 4.96%, N: 22.94%, O: 13.10% and a **formula mass** of 123.11 g mol^{-1}, nicotinamide has the empirical and molecular formula $C_6H_6NO_2$.

Infrared Spectrum

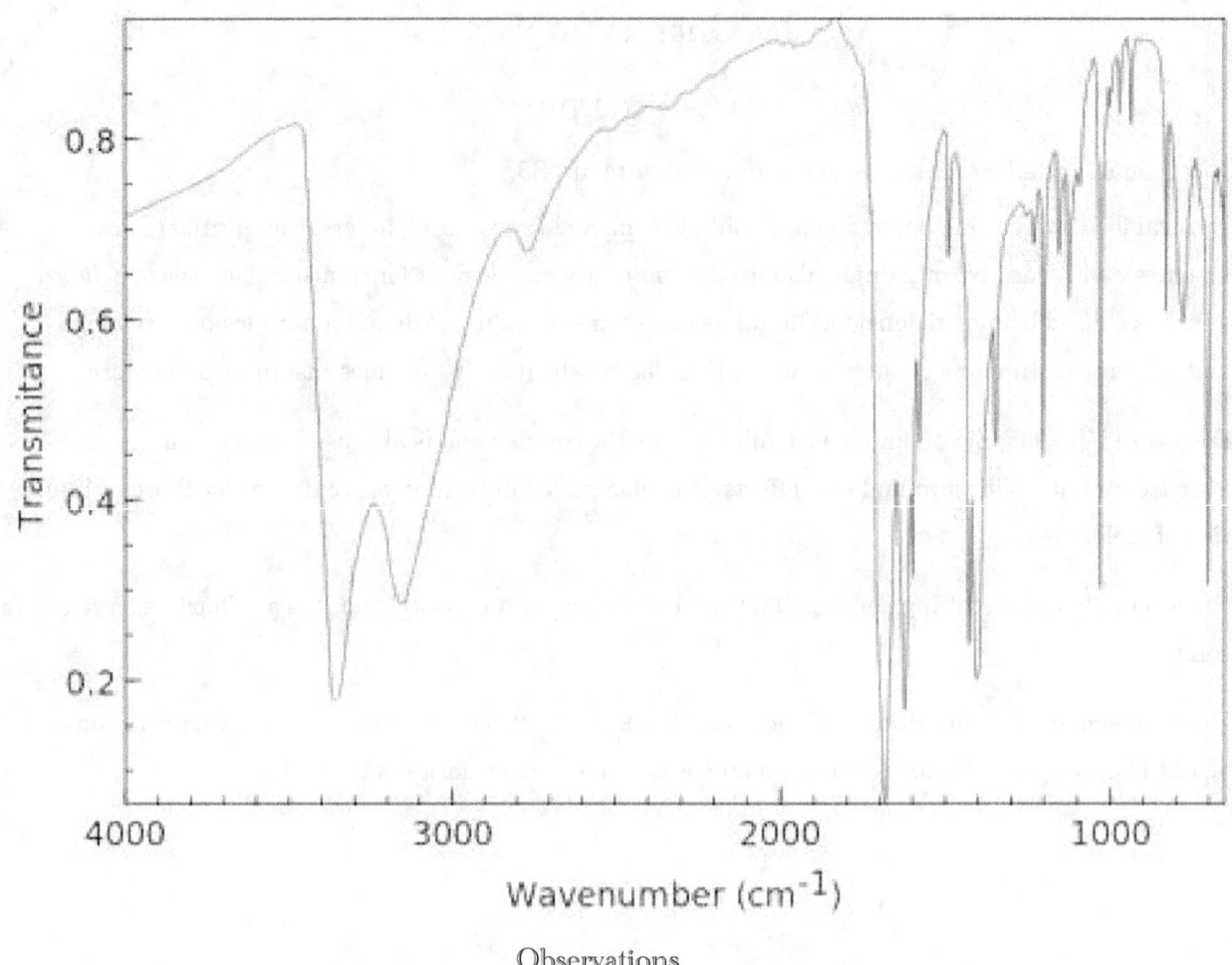

Observations

(√ / X)	Wavenumber range (cm⁻¹)	Wavenumber (cm⁻¹)	Assignment
?	3200 - 3700	3380	O – H
?	3200 - 3600	3380	N – H
√	3000 – 3300	3110	C – H (aromatic)
X	2500 – 3000		C – H (aliphatic)
X	2200 – 2500		C ≡ N
√	1700 – 1800	1690	C = O
X	1600 – 1700		C = C (aliphatic)
X	1585 – 1600		C – C (aromatic)
X	1450 – 1600		C – C (aromatic)
X	1000 – 1300		C – O
X	700 – 1000		C – X (X = Cl, Br or I)

Conclusions

This molecule is clearly aromatic and contains one or more N – H or O – H bonds. Although there is a carbonyl, C=O, bond there is no ester or carboxylic acid functional group since there is no C – O group. Although there is a sharp peak at 1200 cm⁻¹ this is too weak to be due to an ,ester or carboxylic acid, C – O bond.

Mass Spectrum

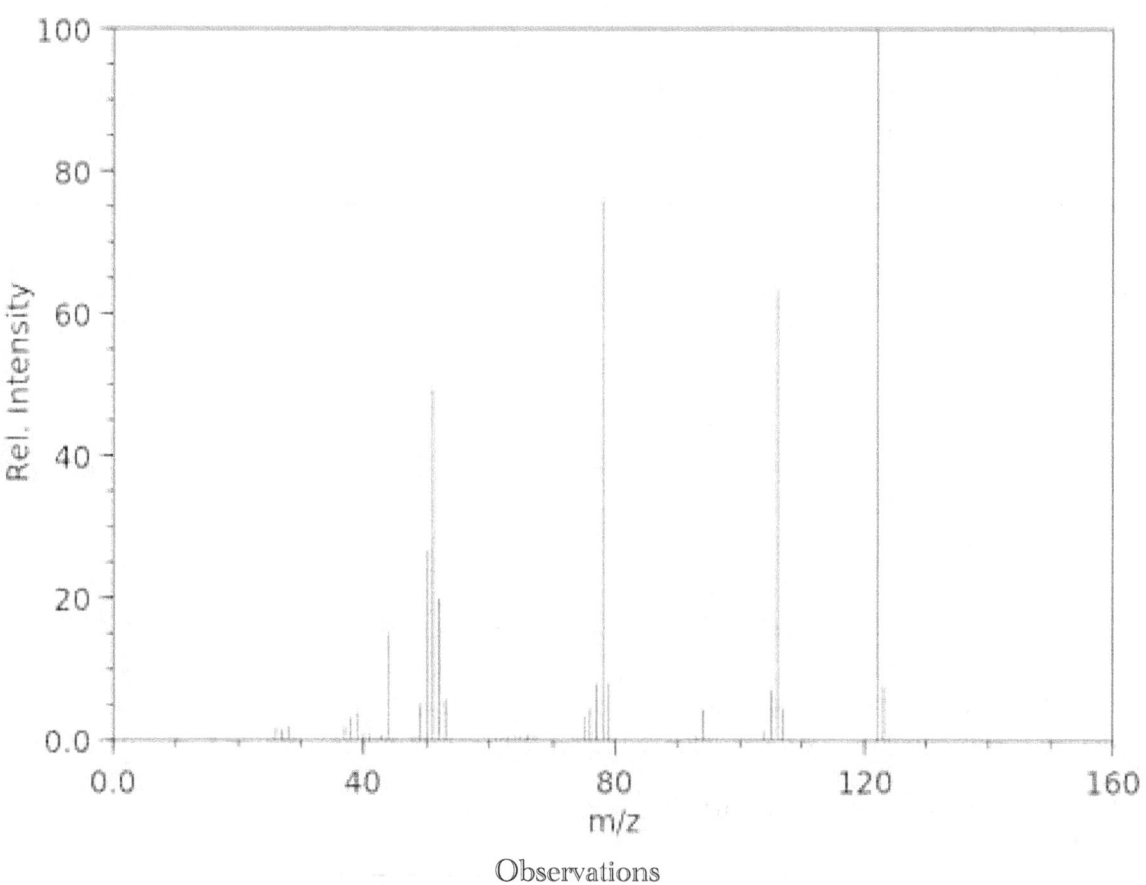

Observations

Charged fragments (m/z)	Assignment	Charged fragments (m/z)	Assignment
Molecular ion: 122	$[C_6H_6N_2O]^+$	Base peak: 122	$[C_6H_6N_2O]^+$
106	$[C_6H_6N_2]^+$	51	$[C_3HN]^+$
78	$[C_6H_5+H]^+$	44	$[CONH_2]+$

Conclusions

The molecule is clearly aromatic but it is difficult to determine how it could contain a benzene ring. An alternative structure could contain a pyridine ring and this would account for the peak at m/z = 78. Not included in the mass spectral data sheet at the beginning of this volume, the fragment could be due to pyridine ring $[C_5H_4N]^+$ which would have one substituent. It is feasible to propose the following structure with one functional group substituent, an amide group, $CONH_2$ which accords with the peak at m/z = 44. We then have the following candidates:

which accords with the infrared spectrum but we can learn much more from the nmr spectra.

NMR Spectra

The ^{1}H and ^{13}C nmr spectra are displayed below:

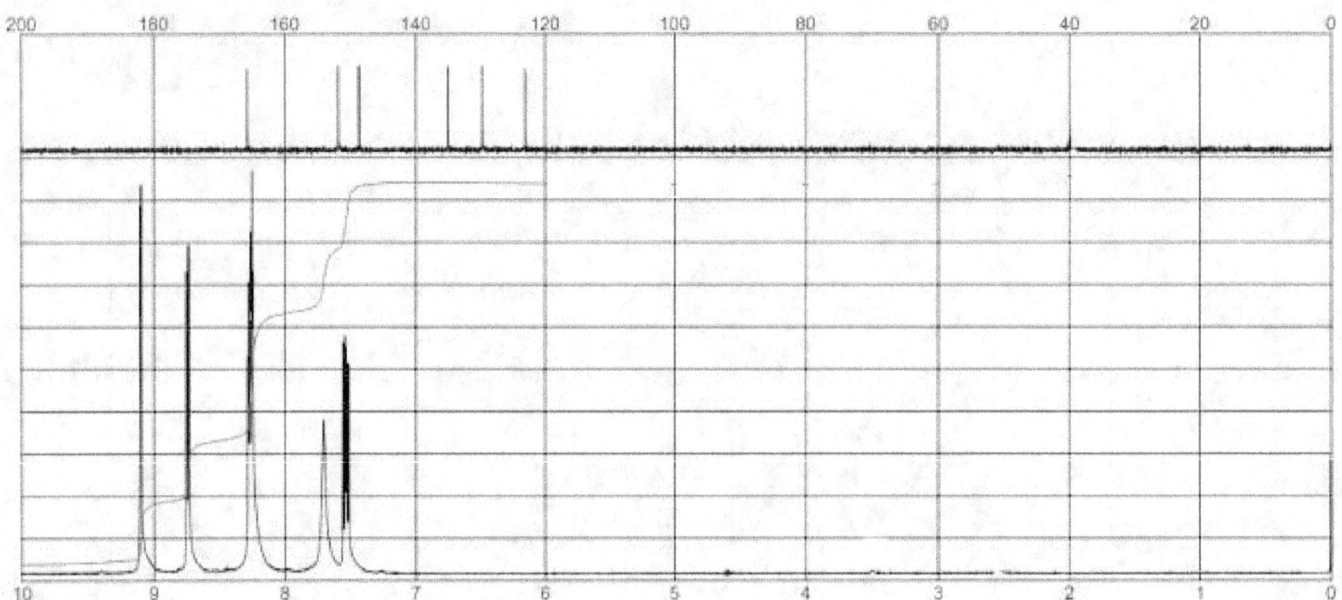

Since the ^{1}H nmr spectrum is complex but the ^{13}C nmr spectra is clear so we will start by considering that.

^{13}C NMR Spectrum

The ^{13}C nmr spectrum is tabulated and partially assigned below.

Chemical shift δ (ppm)	Integral	Assignment
165	1	C = O
153	1	Aromatic
150	1	Aromatic
135	1	Aromatic
130	1	Aromatic
123	1	Aromatic

Observations and Conclusions

There are five aromatic carbon signals which supports the supposition that the aromatic ring is actually a pyridine ring. Since, however, there are five aromatic signals this indicates that the only possible candidates are I and III since candidate II is symetrical.

We know that benzene produces a single ^{13}C nmr resonance signal at δ 128.5 ppm. A small difference may be of no significance but there are two aromatic carbon atoms which are significantly upfield of δ 128.5 ppm. This can only occur due to the presence of an electronegative atom and this, again supports the concept that the molecule contains a pyridine ring with a single substituent as suggested above.

^{13}C carbonyl signals appear within the range δ 160 – 220 ppm. In this case, the signal appears at δ 165 ppm indicating that it is not deshielded and this suggests that Candidate I is the more likely structure.

We can learn more from the ^{1}H nmr spectrum and we will start by predicting the spectrum and then comparing our prediction to the actual observed spectrum

^{1}H NMR Spectrum

The proposed structure is presented again with the hydrogen atoms labelled alphabetically.

The molecule is asymmetric and we can make the following predictions:

- Although the amide grouping can be written as $- C(O)NH_2$, the two amide hydrogen atoms, H_a and H_b, are **not** chemically and magnetically equivalent as the carbonyl, C=O group is planar and so H_a and H_b will produce will each produce doublets of integral one.

- The aromatic hydrogen atom, H_c, will produce a singlet of integral one since there are no hydrogen atoms on adjacent carbon atoms.

- H_d will produce a doublet of doublets of integral one due to sequential splitting by H_e and then H_f.

Hd will produce a doublet of integral one due to splitting by He and then a doublet of doublets due to splitting by Hf.

- H_e will produce a triplet of integral one due to splitting by H_d and H_f.

He will produce a triplet of integral one due to splitting by Hd and Hf.

- H_f will produce a doublet of doublets of integral one due to sequential splitting by H_e and then H_d.

Hf will produce a doublet of integral one due to splitting by He and then a doublet of doublets due to splitting by Hd.

H_c and H_f will be significantly deshielded to their proximity to the electronegative nitrogen atom whilst the doublet of doublets due to H_a and H_b may well overlap.

To summarise we should observe:

Hydrogen atom	Integral	Multiplicity
H_a.	1	Doublet of doublets
H_b	1	Doublet of doublets
H_c	1	Singlet
H_d	1	Doublet of doublets
H_e	1	Triplet
H_f	1	Doublet of doublets

If we examine the expanded ^1H nmr spectrum we observe:

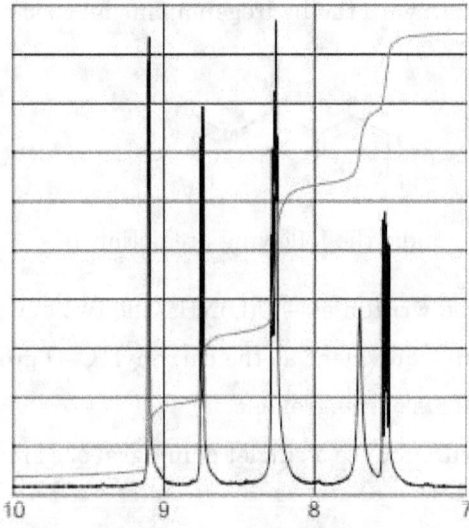

- A singlet of integral one: δ 9.15 ppm, assignable to H_c;
- A doublet of integral one: δ 8.75 ppm; caused by H_f;
- A doublet of doublets of integral one: δ 8.35 ppm, due to H_a or H_b;
- A doublet of doublets of integral one: δ 8.3 ppm, due to H_a or H_b;
- A triplet of integral one: δ 7.75 ppm, due to H_e;
- A doublet of doublets of integral one: δ 7.55 ppm due H_d.

The spectrum confirms the prediction so there can be no doubt about the structure but it is worth noting that, in some molecules analysed in this volume, we have been able to make use of integrals as well as chemical shifts and multiplicities. In this case, the integrals are of little use since they are all one but, on the other hand, they do serve to demonstrate that although the amide hydrogen atoms can be written, in two dimensions, as one group they are subtly different.

Conclusions:

Structure:

Systematic name: pyridine-3-carboxamide

Trivial names: nicotinic acid, vitamin B3

Chapter XV

Quinolin-8-ol

Quinolin-8-ol is a modern treatment for the dreadful disease of malaria.

Malaria, which is is spread by mosquitoes, is a serious tropical disease which causes the deaths of half a million people a year especially in sub-Saharan Africa. It is particularly dangerous for children below the age of five. It is also a disease that was contracted by adventurers to Africa.

The symptoms of malaria range from a high fever to feeling, simultaneously, hot yet shivery as well as headaches, vomiting, diarrhoea and muscle pains. Even when surviving and, apparently better, the symptoms can recur many years later as the infection remains in the blood system.

Quinolin-8-ol has a melting point of -15°C and a boiling point of 237°C and so, at room temperature, it is a colourless crystalline solid which, although sparingly soluble in water, is soluble in ethanol. It can be administered in either ethanolic solution or in tablet form.

Quinolin-8-ol has the elemental composition, C: 74.40%, H: 4.87%, N: 9.69%, O: 11.02% and has a formula mass of 145.16 g mol^{-1}.

This means that the molecule has the **empirical** and **molecular** formula C_9H_7NO.

Infrared Spectrum

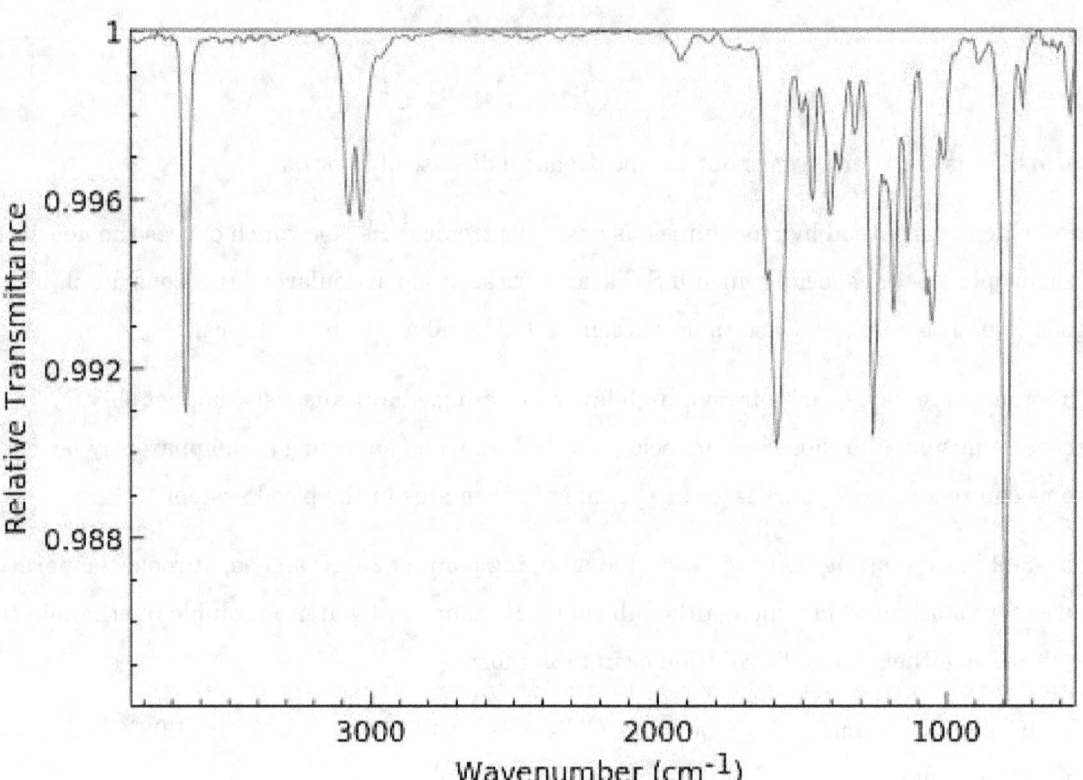

Observations

(√ / X)	Wavenumber range (cm⁻¹)	Wavenumber (cm⁻¹)	Assignment
√	3200 - 3700	3650	O – H
X	3200 - 3600		N – H
√	3000 – 3300	3090, 3010	C – H (aromatic)
X	2500 – 3000		C – H (aliphatic)
X	2200 – 2500		C ≡ N
X	1700 – 1800		C = O
X	1600 – 1700		C = C (aliphatic)
X	1585 – 1600		C – C (aromatic)
√	1450 – 1600	1590	C – C (aromatic)
√	1000 – 1300	1270	C – O
X	700 – 1000		C – X (X = Cl, Br or I)

Conclusions

This compound is aromatic with no aliphatic functional groups or substituents. There must, however, be a hydroxyl group and since there are no aliphatic substituents then this group must be attached directly to the ring.

There is a very strong peak at 795 cm⁻¹ which cannot be assigned as there are no halogens in the compound.

Mass Spectrum

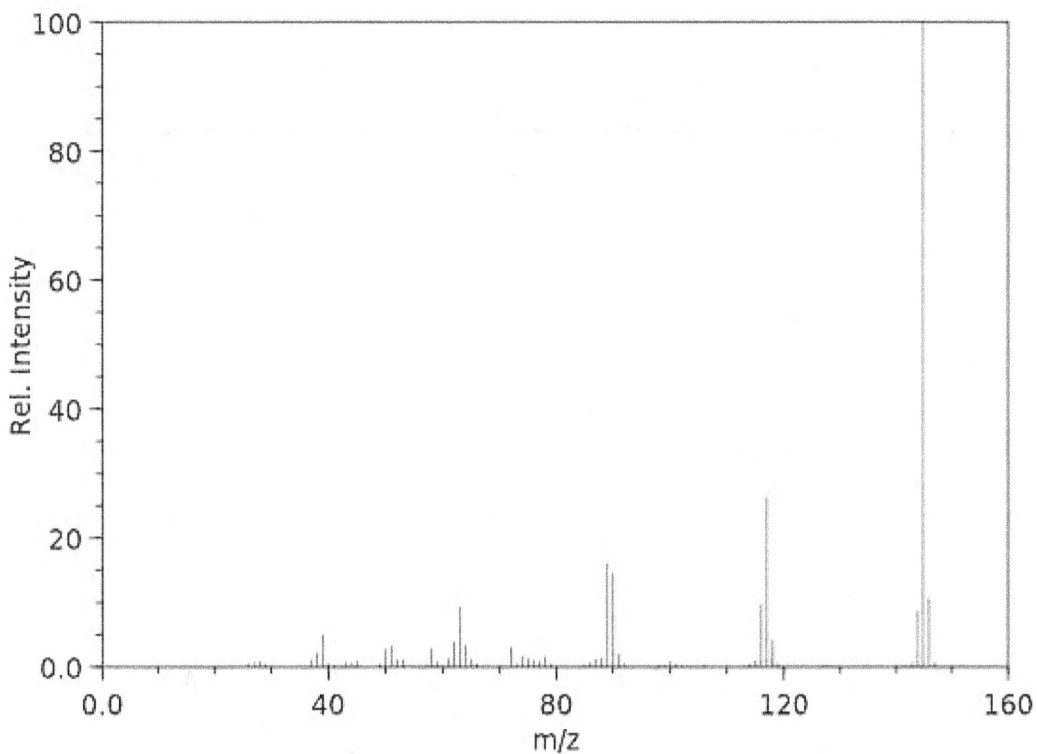

Observations

Charged fragments (m/z)	Assignment	Charged fragments (m/z)	Assignment
Molecular ion: 145	$[C_9H_7NO]^+$	**Base peak: 145**	$[C_9H_7NO]^+$
117	$[C_9H_9]^+$	58	$[C_3H_6O]^+$
90	$[C_7H_6]^+$	39	$[C_3H_3]^+$
63	$[C_5H_3]^+$	29	$[C_2H_5]^+$

Conclusions

We already know that this molecule is entirely aromatic and the presence of the peaks at m/z = 90 and 117 imply the presence of two aromatic rings. The rings must be fused as there are insufficient hydrogen atoms to form an aliphatic bond between two separate benzene rings. In any event, such a structure is prohibited by the absence of any aliphatic C – H stretches in the infrared spectrum.

We must also accommodate the nitrogen atom. There is no N – H peak in the ir spectrum and so this atom must be in the ring itself giving us the following basic structures:

This will leave us to assign an – OH group somewhere on the structure. In principle, the hydroxyl group could appear at any of the aromatic carbon atoms so we have a number of possible molecular structures. There cannot be an N – OH bond as there will be insufficient hydrogen atoms left and the bonding of anything extra to the nitrogen atom will destroy the aromaticity of the fused rings.

We can establish the structure by considering the ^{1}H and ^{13}C nmr spectra which is our next task.

NMR Spectra

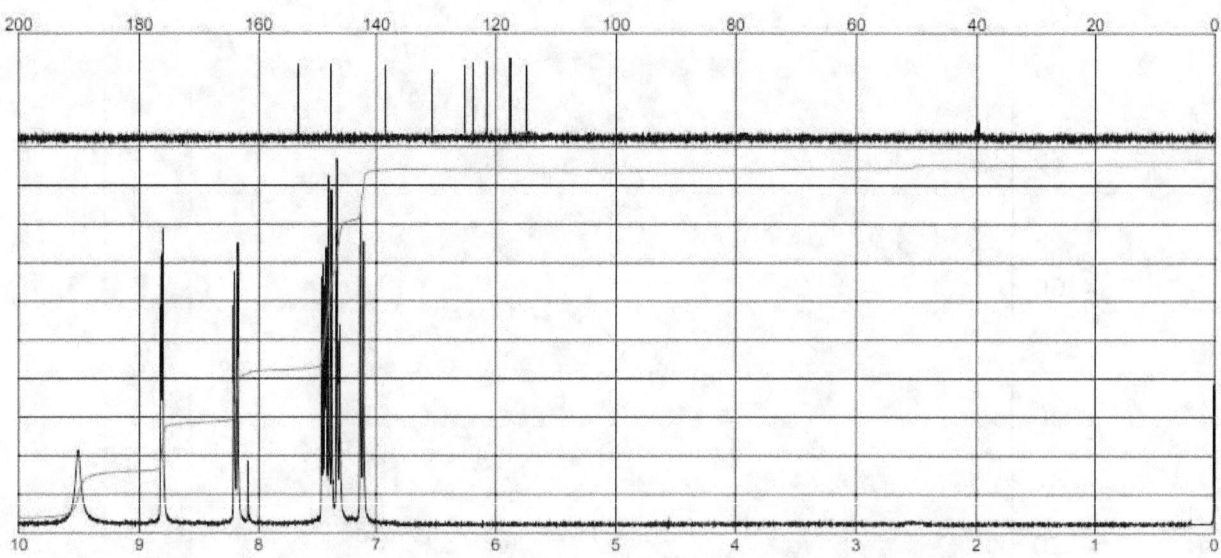

We can make the following, general, assignments from the ^{13}C NMR spectrum

Chemical shift δ (ppm)	Integral	Assignment
154	1	Aromatic
148	1	Aromatic
138	1	Aromatic
130	1	Aromatic
125	1	Aromatic
124	1	Aromatic
122	1	Aromatic
118	1	Aromatic
115	1	Aromatic

This is consistent with the base molecule being a fused benzene and pyridine ring and confirms that there are two possible positions for the nitrogen atom:

I   II

Although the nitrogen atom can be placed in other positions on the ring, rotation or reflection will turn those isomers in to one of the above two possibilities.

We still need to place the – OH group and so we now have the following possibilities:

For **isomer I** the possibilities are:

For **isomer II** the following isomers are possible:

We can immediately discount those molecules where the oxygen and nitrogen atoms are close together on the same ring since, both being highly electronegative, they will deshield the adjacent carbon atoms producing signals of much higher chemical shift than is observed.

This reduces the possible candidates to:

For **isomer I**:

For **isomer II**:

If we examine the ^{13}C nmr spectrum again we observe that five of the signals are very close together and this indicates that five of the carbon atoms in the molecule are similar. This means we can strike out some more of the possible structures as they do not have such a system, leaving us with four possibilities and it will help if we now number the isomers:

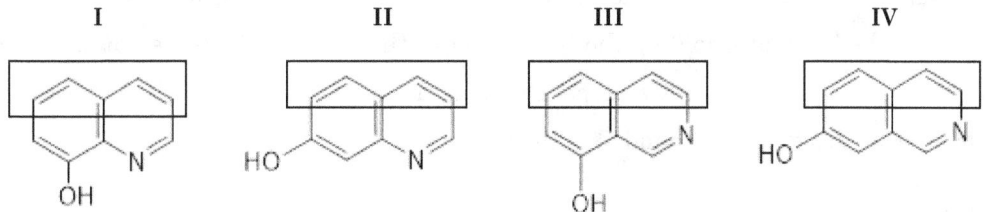

I   II   III   IV

This accounts for five of the signals in the ^{13}C nmr spectrum but do not help us distinguish between them and for this we need to consider the electronegative atoms which can explain the deshielded signals at δ 154, 148 and 138 ppm.

The two highest signals must be due to the carbon atoms bonded to the nitrogen atom and this immediately excludes isomers III and IV as one of the carbon atoms bonded to the nitrogen atom has already been accounted for (the right hand carbon atom in the rectangle highlighted group).

This leaves us with two isomers to choose from:

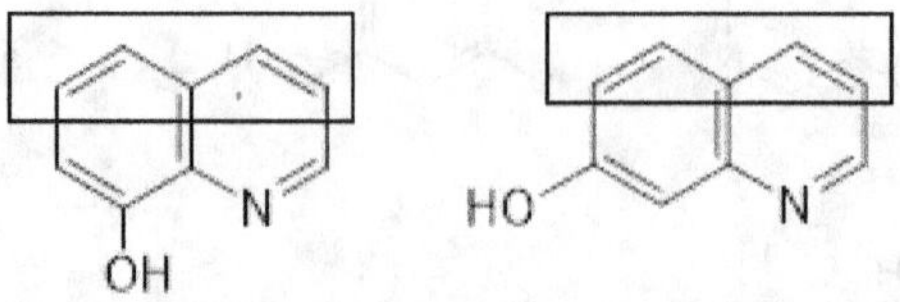

The signal at δ 138 ppm must be due to the carbon atom with the hydroxyl group attached and this simply leaves us with the peak at δ 130 ppm which must be due to the carbon atom highlighted by a circle in the structures below:

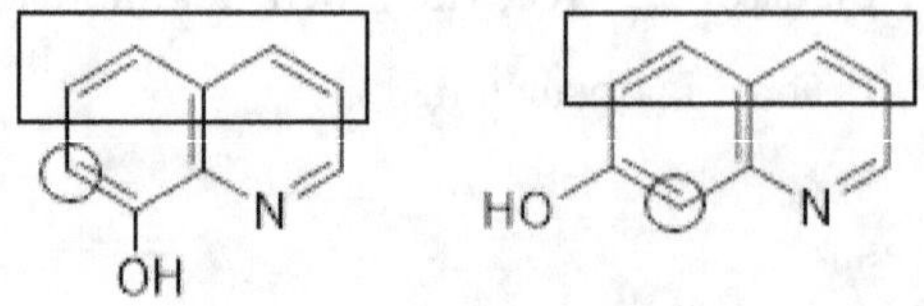

The oxygen atom will deshield the adjacent carbon atoms. If the isomer is structure II then, since the – OH group will deshield the carbon atoms adjacent to the bonding carbon atom, it will deshield the left hand most carbon atom in the highlighted rectangle but this is not observed.

This means that, by a process of elimination, the molecule must be isomer I:

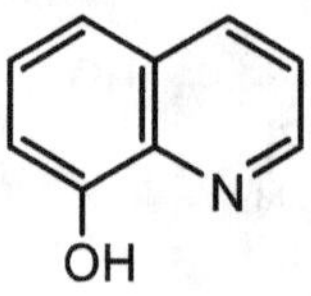

This is, however, not good enough to be conclusive and so we must confirm the structure by examining the ^{1}H nmr spectrum which is our final task. If the ^{1}H spectrum does not support the proposed structure then we must start all over again.

^{1}H NMR Spectrum

Again, this spectrum and the proposed structure appear daunting to analyse but we can relax as the two carbon atoms forming the backbone holding the two rings together have no hydrogen atoms bonded to them and so we can analyse the molecule as two separate rings delineated by the dashed line highlighting the carbons which have no hydrogen atoms bonded to them.

This is shown below:

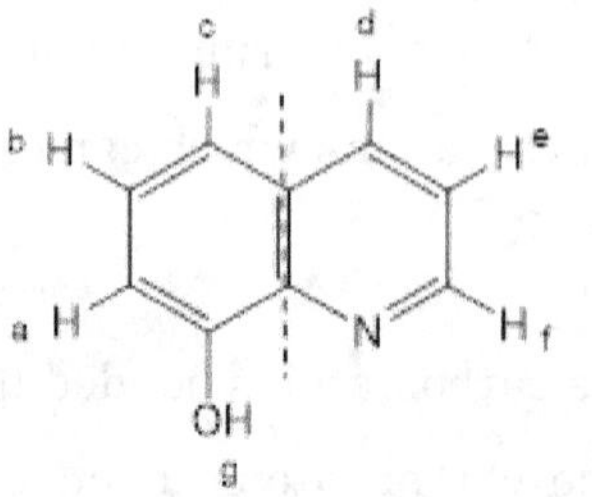

We can easily assign the singlet at δ 9.55 ppm to H_g and we consider H_a, H_b, H_c and H_d, H_e, H_f as two separate groups.

We can make the following predictions:

Isomer I

- H_a will be split into a doublet by H_b which will be split into a doublet of doublets by H_c;
- H_b will be split into a triplet by H_a and H_c;
- H_c will be split into a doublet by H_b and this will be split into a doublet of doublets by H_a.

Isomer II

- H_d will be split into a doublet by H_e which will be split into a doublet of doublets by Hf;
- H_e will be split into a triplet by H_d and Hf;
- Hf will be split into a doublet by H_e and this will be split into a doublet of doublets by H_d.

In detail;

- The doublets of doublet of doublets at δ 8.8 ppm and δ 8.2 ppm can be assigned to H_a and Hf respectively. Since Hf is closer to the electronegative nitrogen than H_a is to the electronegative oxygen atom we can assign Hf to δ 8.8 ppm and H_a to δ 8.2 ppm.
- The multiplet of integral two at δ 7.45 ppm can be assigned to H_c and H_d as they are nearly chemically and magnetically equivalent.

This leaves us with H_b and H_e to assign.

- The triplet at δ 7.10 ppm, of integral one, can be assigned to H_e as it will be affected by the electronegative nitrogen atom which is
- The triplet at δ 7.35 ppm, of integral one, can be assigned to H_b as it will be hardly affected by the presence of the functional, – OH, group and is almost exactly where benzene hydrogen atoms occur.

The assignments are summarised below:

Chemical shift δ (ppm)	Integral	Multiplicity	Assignment
9.55	1	Singlet	H_g
8.80	1	Doublet of doublets	H_a
8.20	1	Doublet of doublets	Hf
7.45	2	Doublet of doublets	H_c / H_d
7.35	1	Triplet	H_e
7.10	1	Triplet	H_b

The observations and conclusions from all four of the infrared, mass, ^{1}H and ^{13}C nmr spectra confirm that the proposed structure is correct.

Conclusions

Structure:

Systematic name: quinolin-8-ol

Trivial names: 8-hydroxyquinoline, 8-quinolinol, oxyquinoline

Chapter XVI

Methyl nicotinate

Methyl nicotinate is closely related to nicotinic acid (niacin, Vitamin B3) and menthol.

It is used in cosmetics, personal care products and as, over the counter remedies for athletic injuries. It is also used in veterinary purposes to treat respiratory disease, vascular disorders, rheumatoid arthritis as well as, again, muscle and joint pains.

This substance has a melting point of 40°C and a boiling point of 204°C and has the elemental composition: C: 61.25%, H: 5.16%, N: 10.21%, O 23.33% With a formula mass of 137.14 g mol^{-1} this means that both the *empirical* and *molecular* formulas are $C_7H_7NO_2$.

The melting point of 40°C means that the compound is a colourless, crystalline solid.

The molecule possesses two oxygen atoms which immediately suggests that it might be a carboxylic acid or an ester. It cannot be a diol (a molecule with two – OH groups) due to there being insufficient number of remaining hydrogen atoms to complete the structure.

It is fairly soluble in water (48 g dm^{-3} at 20°C) and that means that it can be administered in either tablet form or as an aqueous solution. Methyl nicotinate is highly soluble in common organic solvents such as chloroform, methanol and ethyl ethanoate (commonly known as ethyl acetate). This property is unlikely for a carboxylic acid and so it is most likely to be an ester.

Infrared Spectrum

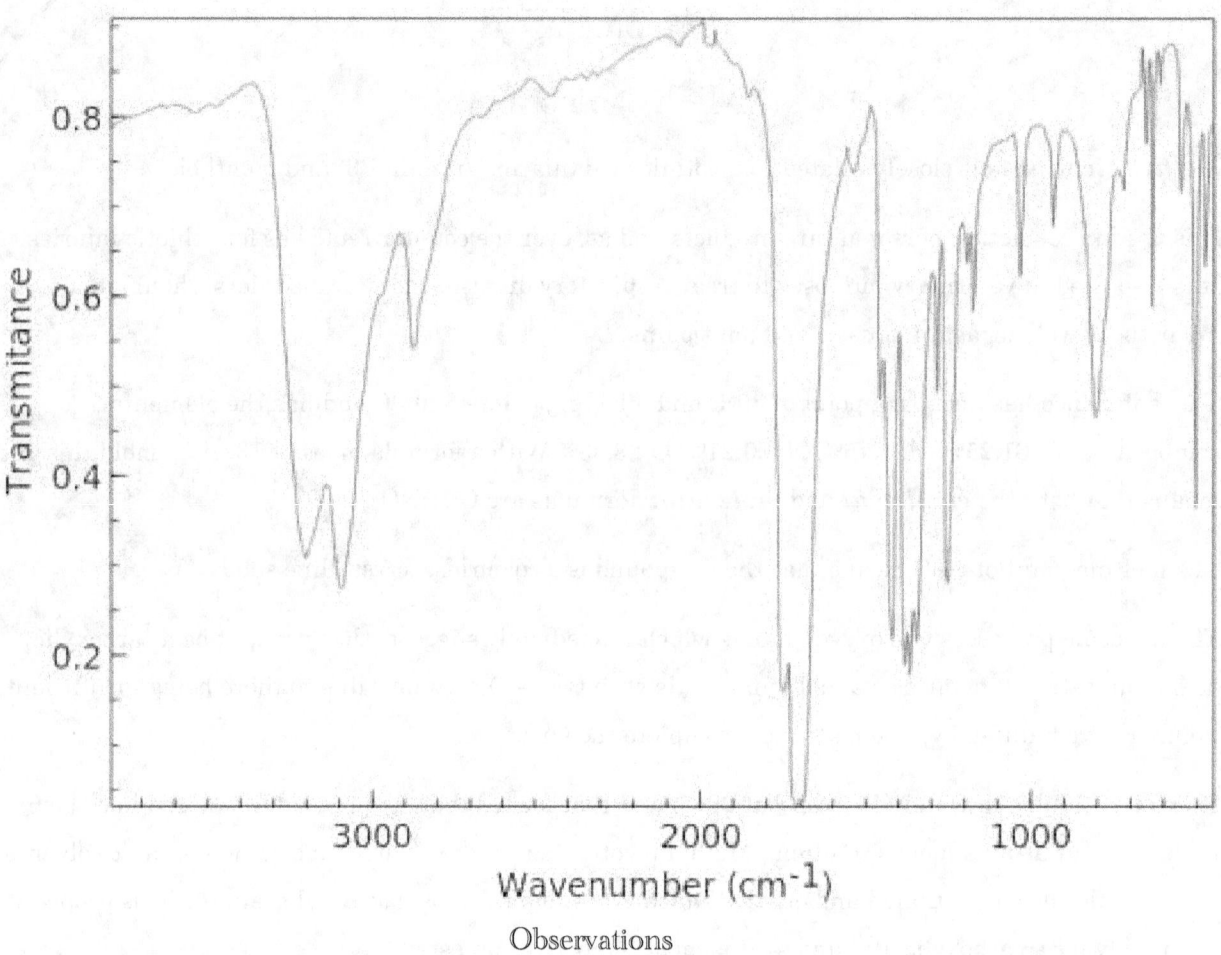

Observations

(√ / X)	Wavenumber range (cm⁻¹)	Wavenumber (cm⁻¹)	Assignment
X	3200 - 3700		O – H
X	3200 - 3600		N – H
√	3000 – 3300	3200, 3080	C – H (aromatic)
√	2500 – 3000	2850	C – H (aliphatic)
X	2200 – 2500		C ≡ N
√	1700 – 1800	1710	C = O
X	1600 – 1700		C = C (aliphatic)
X	1585 – 1600		C – C (aromatic)
X	1450 – 1600		C – C (aromatic)
√	1000 – 1300	1250	C – O
X	700 – 1000		C – X (X = Cl, Br or I)

Conclusions

This compound is:-

- Aromatic with an aliphatic substituent which must contain an ester functional group (ester groups always appear as a terminal group).
- It cannot be a carboxylic acid as there is no evidence for an O – H group.

Mass Spectrum

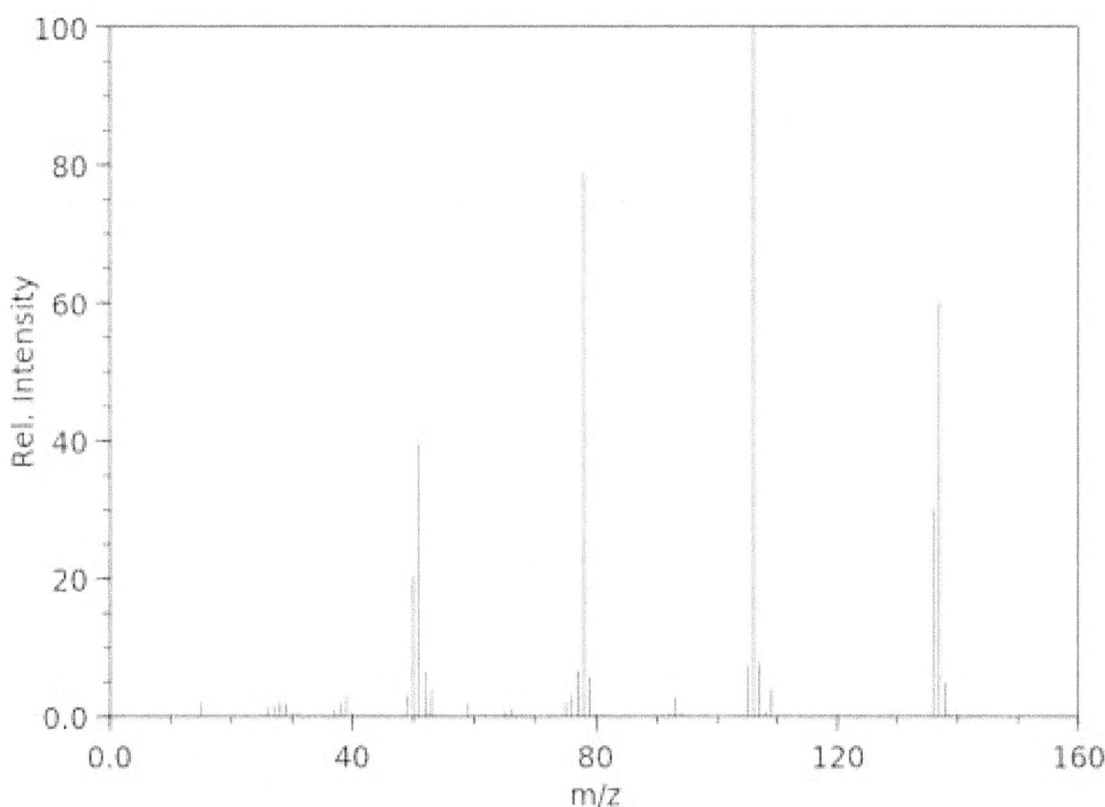

Observations

Charged fragments (m/z)	Assignment	Charged fragments (m/z)	Assignment
Molecular ion: 137	$[C_7H_7NO_2]^+$	**Base peak: 106**	$[C_7H_6O]^+$

93	$[C_6H_5O]^+$	51	$[C_4H_3]^+$
78	$[C_5NH_4]^+$	39	$[C_3H_3]^+$
59	$[C_3H_7O]^+$	15	$[CH_3]^+$

Conclusions

The most significant peak is that at m/z = 78 which implies the existence of of a pyridine ring with one substituent. there are three possible isomers, where the squiggly line represent the sole functional group:

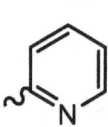 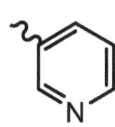 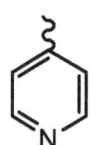

The substituent has the m/z value of 59 which indicates that it has the formula $C_2H_3O_2$. This could only be due to a $C(O) - OCH_3$ group and we can establish the exact isomeric structure from the ^1H and ^{13}C nmr spectra. This is our next task.

NMR Spectra

The ^{1}H and ^{13}C nmr spectra are displayed below:

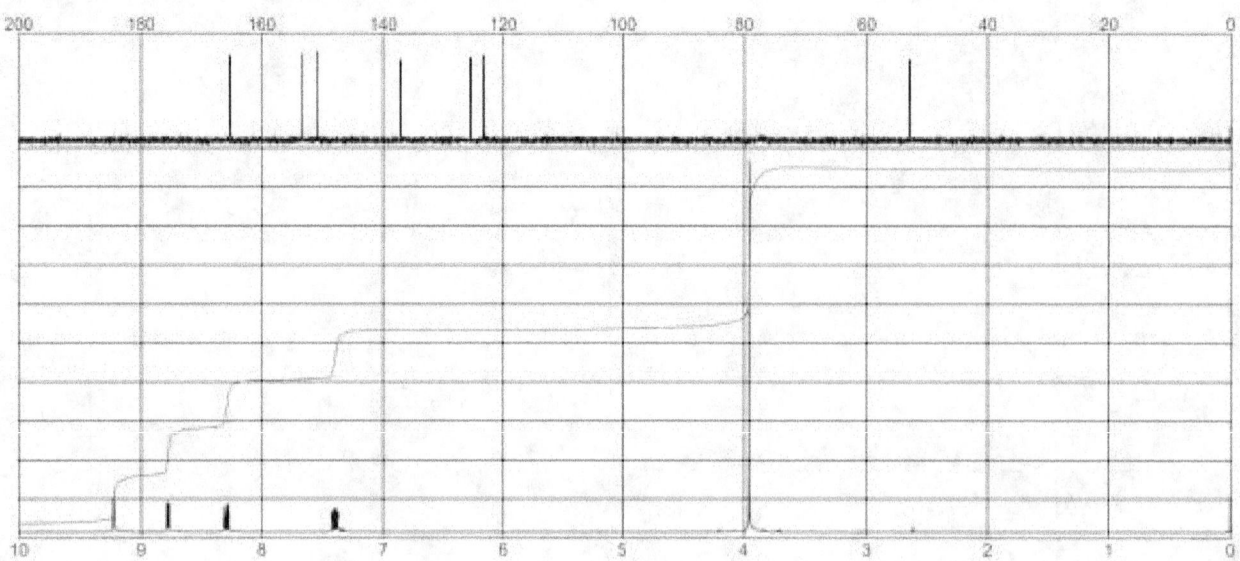

If the proposed structure is correct i.e. a pyridine ring with a sole C_2H_3O functional group then we have three possible isomeric structures:

I	II	III

Before examining the actual ^{1}H and ^{13}C nmr spectra we can predict the chemical shifts and multiplicities that we could observe.

^{13}C NMR Spectrum Predictions

- For isomers I and II we will observe:
    - Five signals assignable to the carbon atoms in the pyridine ring in the region δ 110 – 160 ppm;
    - A singlet of integral one due to the carbonyl atom within the ester grouping in the region δ 160 – 220 ppm range;
    - A singlet due to the $CH_3 – O –$ terminal carbon atom, δ 50 – 90 ppm.
- Isomer III is symmetrical so we should observe three peaks, two of integral two and one of integral one with the substituent. Since there are five singlets, all of integral one, isomer III cannot be the structure of the molecule.

This leaves us with isomers I and II so our next, relatively straightforward task, is to predict the ^{1}H nmr spectra for both isomers.

Predicting the ^{1}H NMR Spectrum

The ^{1}H nmr spectra will be considered for each isomer in turn with the hydrogen atoms labelled alphabetically.

Isomer I:

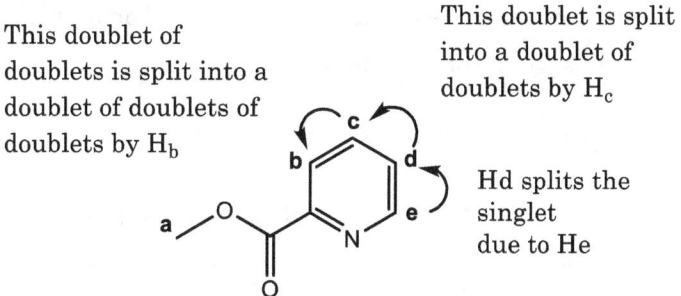

- H_a will appear as a singlet of integral three in the range δ 3 – 4.5 ppm;

- H_b will produce a doublet of doublet of doublets due to sequential splitting of the singlet by H_c, H_d and then H_e:

H_g is split into a doublet by H_h and into a doublet of doublets by H_j

H_j produces a singlet

- H_c will produce a doublet of doublets of doublets due to sequential splitting of the original singlet H_c by H_d and then by H_e.

H_b splits H_c into a doublet

This doublet is split into a doublet of doublets by H_d

This doublet of doublets is split into a doublet of doublets of doublets by H_e

- H_d will produce a doublet of doublet of doublets due to splitting by H_e and then, sequentially by H_c and H_b.

This doublet of doublets is split into a doublet of doublets of doublets by H_b

This doublet is split into a doublet of doublets by H_c

Hd splits the singlet due to He

In summary then we would observe:

- A singlet due to the alkyl hydrogen atom;

- One doublet of doublets;

- Two doublets of doublets of doublets.

Isomer II:

- H_j will produce a singlet of integral one as there are no hydrogen atoms on adjacent carbon atoms.

- H_g will produce a doublet of doublets since its singlet is split into a doublet by H_h and then into a doublet of doublets by H_j.

H_g is split into a doublet by H_h and into a doublet of doublets by H_j

H_j produces a singlet

- H_h will produce a triplet of integral one due to splitting by H_g and H_i:

H_h is split into a triplet by H_g and H_j

H_j produces a singlet

- H_i will produce a doublet of doublets due to sequential splitting by H_h and then H_g:

H_i is split into a doublet by H_h and into a doublet of doublets by H_g

H_j produces a singlet

In summary then we would observe:

- A singlet due to H_j
- A triplet due to H_h
- Two doublets of doublets assignable to H_g and H_i.

The doublets due to Hi and Hj will be significantly deshielded due to the carbon atoms being adjacent to the electronegative nitrogen atom. All signals will be of integral one.

Comparing the two sets of predictions

It is helpful to summarise the predictions, alkyl and aromatic, before examining the actual spectrum.

Isomer	Alkyl	Aromatic
I	Singlet of integral three	- One singlet of integral one - One doublet of doublets, of integral one - Two doublets of doublets of doublets each of integral one
II	Singlet of integral three	- One singlet of integral one - One triplet of integral one - Two doublets of doublets each of integral one

The alkyl singlet will be present in the spectra of all three isomers so we need only consider the aromatic region which is shown in expanded form below:

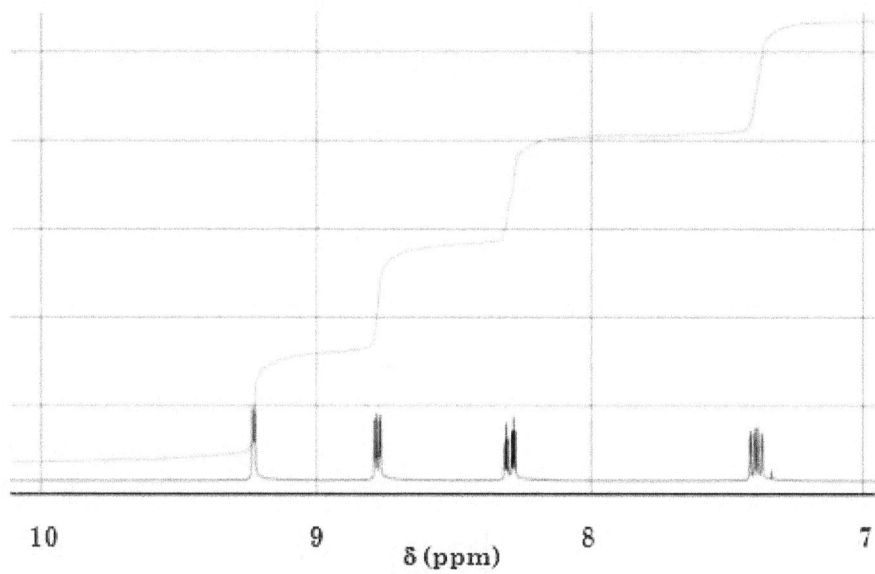

It is clear that the aromatic region of the spectrum comprises one singlet of integral one, two doublets of integral two and a triplet also of integral one.

This means that only isomer II is a possible candidate and we can make the following assignments on the basis of this structure:

Chemical shift δ (ppm)	Integral	Multiplicity	Assignment
9.30	1	Singlet	H_j
8.80	1	Doublet of doublets	H_i
8.35	1	Triplet	H_h
7.40	1	Doublet of doublets	H_g
3.98	3	Singlet	H_f

but to finally confirm this we must review the ^{13}C nmr spectrum again.

^{13}C NMR Spectrum

We have already observed that the ^{13}C nmr spectrum contains:

- Five signals assignable to the carbon atoms in the pyridine ring in the region δ 110 – 160 ppm;
- A singlet of integral one due to the carbonyl atom within the ester grouping in the region δ 160 – 220 ppm range;
- A singlet due to the $CH_3 – O –$ terminal carbon atom, δ 50 – 90 ppm.

These are all observed but we can examine it in more detail with reference to the proposed structure shown below with new alphabetical labelling:

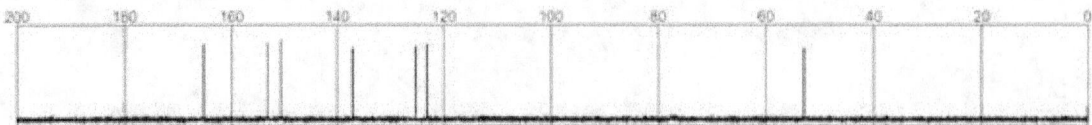

There are five peaks as shown below:

Two peaks (δ 164 ppm and δ 54 ppm) can be readily assigned to Cp and Co respectively and this leaves with the five peaks in the region δ 120 – 160 ppm. These are all due to the five carbon atoms in the ring. The ring is asymmetrical and this explains the existence of the five peaks.

Two of the peaks (δ 151 and δ 154 ppm) are significantly deshielded and are very close to each other and these can be assigned to C_t and C_u as they will be deshielded by the electronegative nitrogen atom. It is not possible with the information available to determine the exact assignments though.

This leaves us with C_q, C_r and C_s to assign. C_q and C_s are adjacent to C_u and C_t respectively and we can plausibly assign the peaks at δ 124 ppm and δ 122 ppm which leaves the peak at δ 137 ppm to be assigned to C_r.

Conclusions

Structure:

Systematic name: methyl nicotinate

The naming is based on nicotinic acid which is another form of Vitamin B$_3$ and has the structure:

and this molecule is the methyl ester.

Chapter XVII

Pyrazinamide

Synthesised in 1934, ***pyrazinamide*** has, since 1952, been used as an oral treatment for tuberculosis in combination with medications such as rifampicin, isoniazid and ethambutol which is the standard four drug regimen. It is never used on its own and is generally used only in the first two months of treatment to reduce the duration of treatment. Other treatments, not including pyrazinamide, typically require at least nine months of treatment.

It has many side effects including minor ones such as nausea, rashes, loss of appetite, muscle and joint pains, and rash whilst more significant side effects include gout and liver damage.

It is not given to patients also suffering from porphyria or serious liver disease.

Pyrazinamide is a colourless white solid with a melting point of 192°C and a boiling point of 357°C.

It has an elemental composition C: 48.74%, H: 4.10%, N: 34.16%, O: 13.00% and has a formula mass of 123.11 g mol^{-1} making both the **empirical** and **molecular** formulas $C_5H_5N_3O$.

Infrared Spectrum

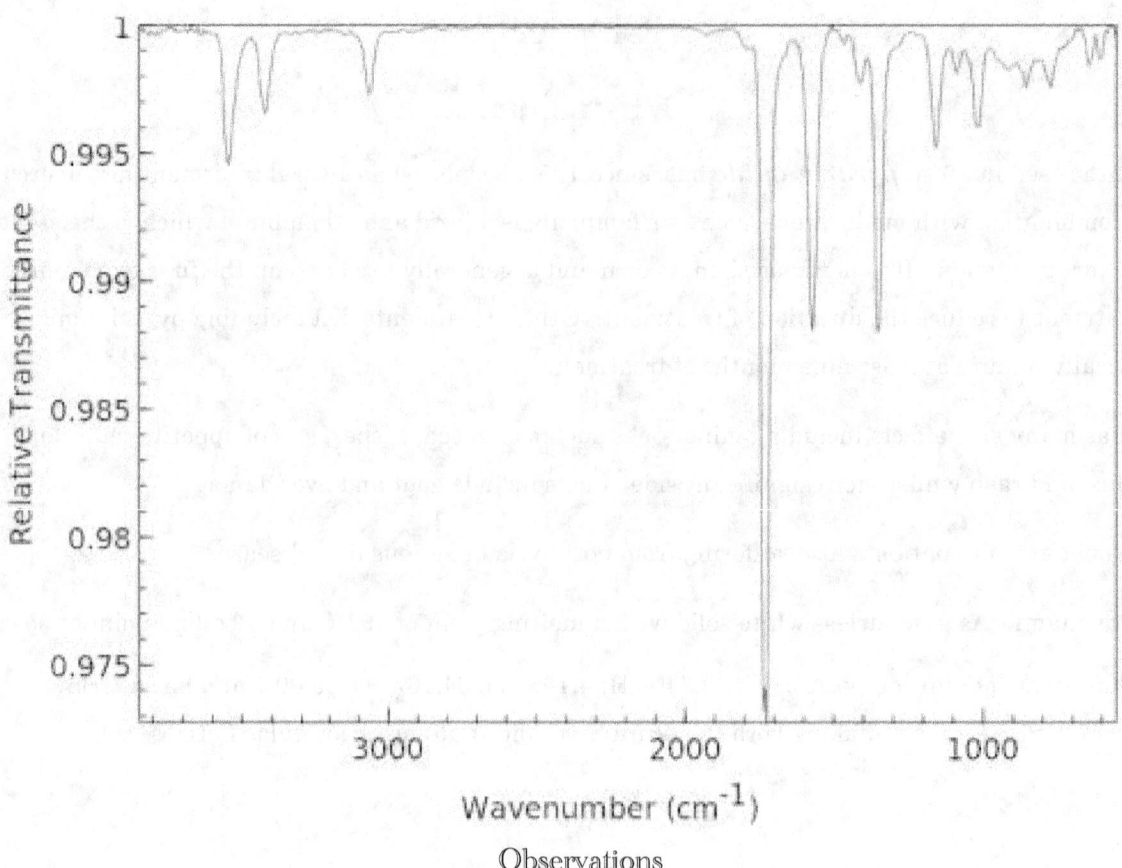

Observations

(√ / X)	Wavenumber range (cm⁻¹)	Wavenumber (cm⁻¹)	Assignment
X	3200 - 3700		**O – H**
√	3200 - 3600	3650, 3460	**N – H**
√	3000 – 3300	3020	**C – H (aromatic)**
X	2500 – 3000		**C – H (aliphatic)**
X	2200 – 2500		**C ≡ N**
√	1700 – 1800	1720	**C = O**
X	1600 – 1700		**C = C (aliphatic)**
X	1585 – 1600		**C – C (aromatic)**
√	1450 – 1600	1550	**C – C (aromatic)**
X	1000 – 1300		**C – O**
X	700 – 1000		**C – X** (X = Cl, Br or I)

Conclusions

- This molecule is aromatic and contains both an N – H and C = O group;

- There is no O – H group so this molecule cannot be a carboxylic acid;

- This is confirmed by the absence of any C – O bond;

- The absence of any aliphatic C – H groups also removes the possibility of the molecule being an ester;

- The presence of three nitrogen atoms suggests that there is at least one nitrogen in the aromatic ring.

Mass Spectrum

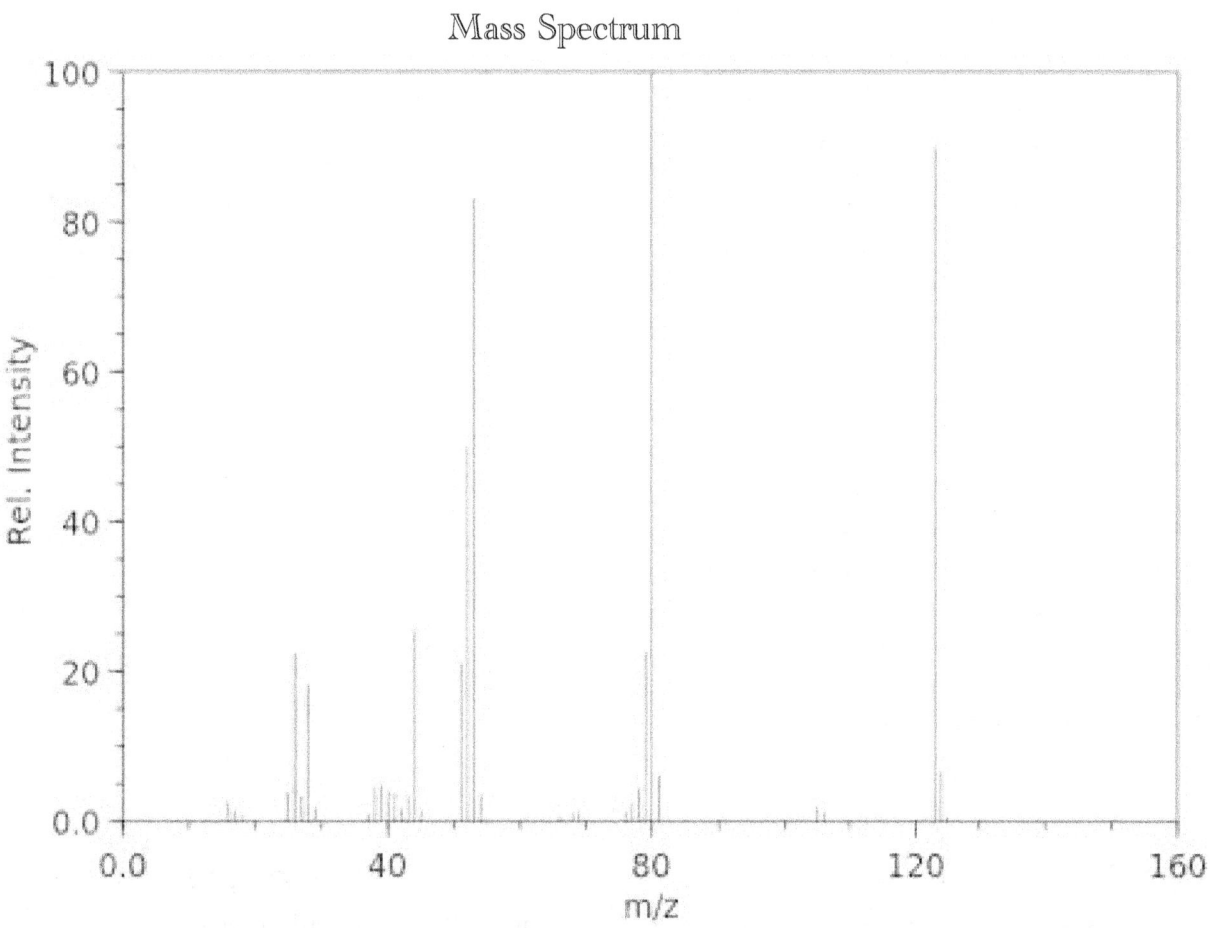

Observations

Charged fragments (m/z)	Assignment	Charged fragments (m/z)	Assignment
Molecular ion: 123	$[C_5H_5N_3O]^+$	Base peak: 80	$[C_5H_4O]^+$

53	$[C_4H_5]^+$	28	$[C_2H_4]^+$
44	$[CONH_2]^+$	26	$[C_2H_2]^+$

Conclusions

The peak at m/z = 44 equates to the functional group

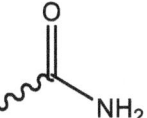

where the squiggly line indicate an undetermined bond.

This implies that there is a ring to which this functional group is attached which must be aromatic and contain four carbon atoms and two nitrogen atoms.

If there is an aromatic group with this functional group then we can learn far more from the ^{1}H and ^{13}C nmr spectra which is our next task.

NMR Spectra

The ^1H and ^{13}C nmr spectra are displayed below:

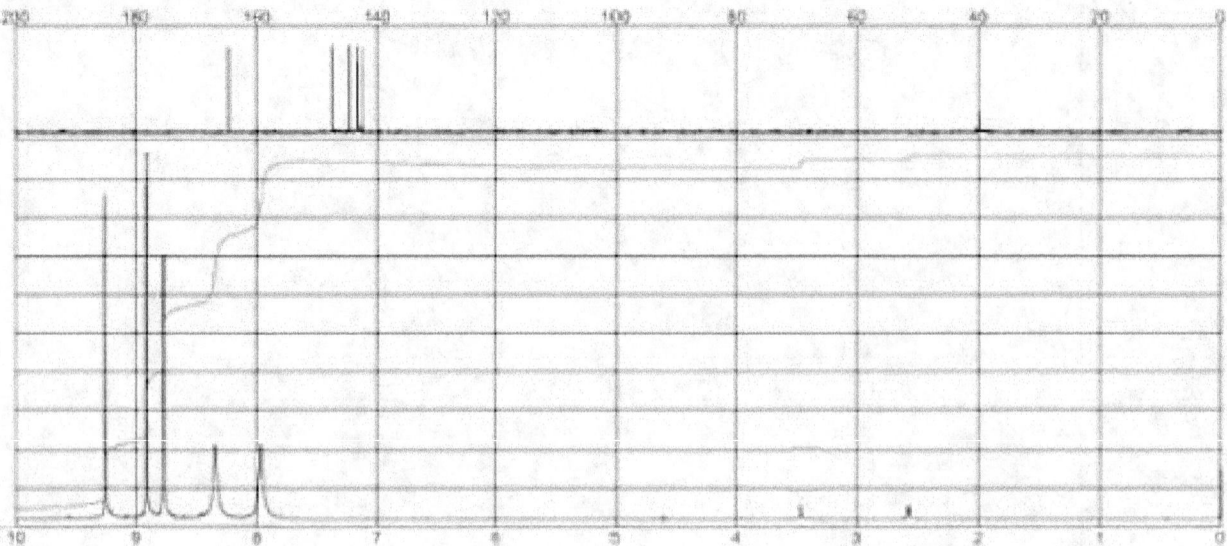

We will start with the ^{13}C nmr spectrum.

^{13}C NMR Spectrum

The peak at δ 165 ppm, of integral one, can be readily assigned to a carbonyl group whilst the four peaks, all of integral one, at δ 147 ppm, δ 145 ppm, δ 143 ppm and δ 142 ppm are all in the aromatic region which makes sense as there are too few hydrogen atoms for a linear molecule.

Since there are only four aromatic carbon atoms the remaining two places, if the aromatic molecule contains a six – membered ring, must both be nitrogen atoms leading to the possibility of the following aromatic structure:

where the squiggle represents the substituted group

The formula of this fragment is $C_4H_3N_2$ which leaves the functional group to be CH_2O. This can only be of the form:

There can only be one isomer since wherever the substituted group is placed, simple rotation turns alternative isomers into the one shown below:

This structure accords with the ^{13}C nmr spectrum and, labelling the carbon atoms, we can assign the peaks as follows:

- The signal at δ 165 ppm is clearly due to C_e;
- C_a and C_b are nearly chemically and magnetically and will be subject to the same amount of deshielding so we can assign the peaks at δ 142 and δ 143 ppm to these two carbon atoms. It is not, however, possible to exactly assign the specific signals;
- C_c will also be deshielded by being adjacent to a nitrogen atom and can be assigned to δ 145 ppm;
- C_d will be deshielded by a nitrogen atom and also by Ce which is, itself, also deshielded by both an oxygen and a nitrogen atom and can be assigned to the δ 147 ppm signal.

These can be summarised as follows:

Chemical shift δ (ppm)	Integral	Assignment
147	1	C_d
145	1	C_e
143	1	C_a or C_b
142	1	C_a or C_b
165	1	C_e

This all looks promising but we cannot assign the structure without analysing the ^1H nmr spectrum and we will start by predicting the spectrum. If at any point the prediction does not match the observed spectrum then the prediction is incorrect as the ^1H nmr spectral measurements are not incorrect.

^1H NMR Spectrum

It is clear that the singlet at δ 3.48 ppm and the doublet at δ 2.58 ppm can be ignored since they are both of much less than integral one and can be assumed to be impurities.

Another essential aspect to note is that since the carbonyl, C=O, bond is planar then, although we could draw the – NH_2 as one group, the planarity of the carbonyl group means that the two amino hydrogen atoms are not chemically and magnetically equivalent. It is important to also remember that nitrogen is tetrahedral due to its lone pair.

We can relabel the molecule, alphabetically starting at f to ensure there is no confusion and then predict the spectrum. Using the new labelling we can describe the proposed structure as shown below:

We can make the following predictions:

- H_f will be a doublet of integral one due to splitting by H_g as shown below:

H_f is split into a doublet by H_g

and will appear above δ 6 ppm;

- H_g will be a doublet of integral one due to splitting by H_f:

H_g is split into a doublet by H_f

and will also appear above δ 6 ppm.

- H_h will be a singlet of integral one, again as there are no hydrogen atoms on the adjacent carbon atom and will appear above δ 6 ppm;

- H_i and H_j will both produce doublets of integral one.

If we consider the ^1H nmr spectrum for the region we observe five signals all of integral one, two are doublets (δ 8.90 ppm and δ 8.75 ppm), one sharp singlet at δ 9.30 ppm and two broad singlets at δ 8.35 ppm and δ 7.96 ppm which are identical and so must be due to amino hydrogen atoms. We can make the following assignments using the labelling above:

Chemical shift δ (ppm)	Integral	Assignment
9.30	1	H_h
8.90	1	H_f or H_g
8.75	1	H_f or H_g
8.35	1	H_i or H_j
7.96	1	H_i or H_j

It is not plausible to confidently assign specific peaks to H_f and H_g or to H_i and H_j but this shows the power of using the multiplicities and integrals to determine the structure and we can be confident that the structure is as predicted as it complies with the infrared, ^1H and ^{13}C nmr spectra and, to a lesser extent, the mass spectrum which demonstrated the existence of the $- C(O)NH_2$ functional group.

Conclusions

Structure:

Systematic name: pyrazine-2-carboxamide.

Chapter XVIII

Clioquinol

Clioquinol is used as an anti-bacterial and anti-fungal constituent of many medicines, addressing skin disorders.

It is also used as treatment for prostrate cancer due to its interaction with zinc but that mechanism is not well understood and, in any event, would be well outside of the scope of this volume. Some studies have suggested that clioquinol can halt the progress of all of Alzheimer's, Parkinson's and Huntington's conditions but, 2021, that has yet to be proven.

Since clioquinol has a melting point of 179°C and a boiling point of 350°C it is, at room temperature, a colourless, crystalline solid.

Clioquinol has the elemental composition: C: 35.35%, H: 1.65%, N: 4.58%, O:5.24%, I: 41.54% and has the formula mass (M_r) of 305.50 g mol^{-1}.

This means that the **empirical** and **molecular** formulas are both C_9H_5NOClI.

Infrared Spectrum

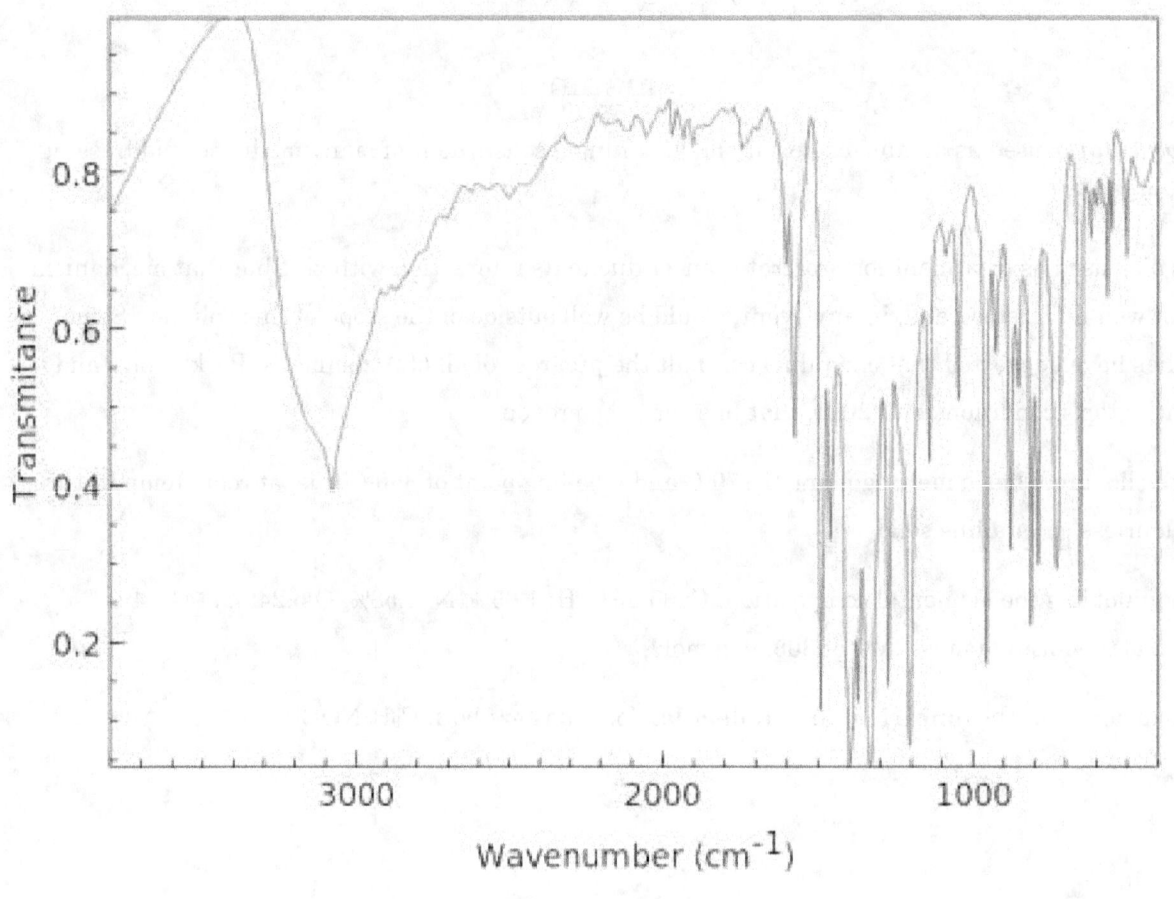

Wavenumber (cm^{-1})

Observations

(√ / X)	Wavenumber range (cm^{-1})	Wavenumber (cm^{-1})	Assignment
√	3200 - 3700	3220	O – H
X	3200 - 3600		N – H
√	3000 – 3300	3050	C – H (aromatic)
X	2500 – 3000		C – H (aliphatic)
X	2200 – 2500		C ≡ N
X	1700 – 1800		C = O
X	1600 – 1700		C = C (aliphatic)
√	1585 – 1600	1595	C – C (aromatic)
√	1450 – 1600	1500	C – C (aromatic)
X	1000 – 1300		C – O
√	700 – 1000	960, 820	C – X (X = Cl, Br or I)

Conclusions

▣ This spectrum is extremely unusual in the shape of the stretch to the left of 3000 cm^{-1} and it must be due to an O – H stretch.

▣ More interestingly there are no peaks in the aliphatic (2500 – 3000 cm^{-1}) region and so this molecule must be aromatic only.

▣ There are peaks due to C – I and C – Cl bonds.

Mass Spectrum

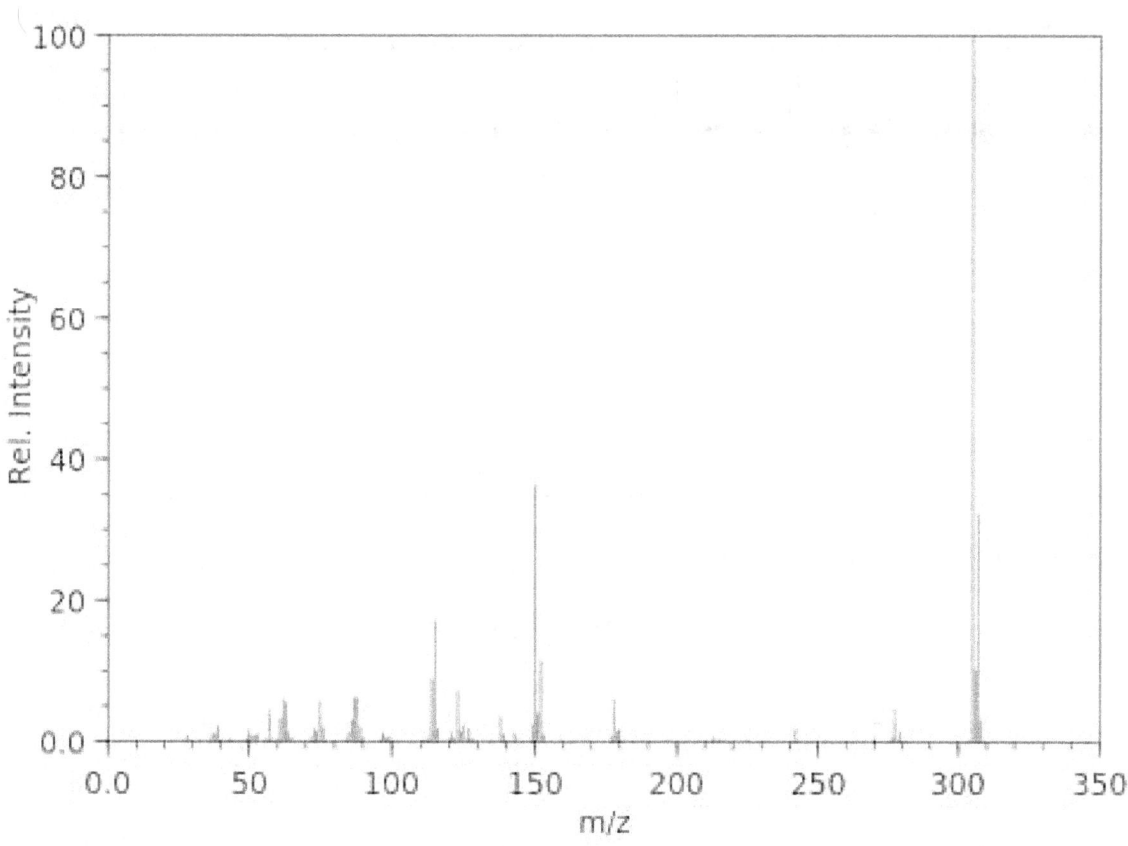

Observations

Charged fragments (m/z)	Assignment	Charged fragments (m/z)	Assignment
Molecular ion: 305	$[C_9H_5NOClI]^+$	Base peak: 305	$[C_9H_5NOClI]^+$
277	$[C_8H_5NClI]^+$	115	$[C_8H_5N]^+$
242	$[C_8H_5NI]^+$	88	$[C_7H_4]^+$
178	$[C_9H_5NOCl]^+$	75	$[C_6H_3]^+$
150	$[C_8H_5NCl]^+$	62	$[C_5H_2]^+$
123	$[C_7H_4Cl]^+$	57	$[C_4H_9]^+$

Conclusions

This spectrum gives us some but limited useful information beyond confirming the formula mass of the molecule. It is, however, interesting and, perhaps very significant that the molecular ion and the base peak are the same indicating that the molecule is stable even when ionised. This supports the possibility that the molecule is stable and aromatic.

- The fragmentation reveals no evidence for the presence of an aliphatic functional group and the fragmented peaks may be due to fragmentation of the ring;

- The peaks at m/z = 150 & 152 and m/z = 305 & 307 are in the ratio 3:1 indicating that these two signals describe a fragment continuing chlorine.

NMR Spectra

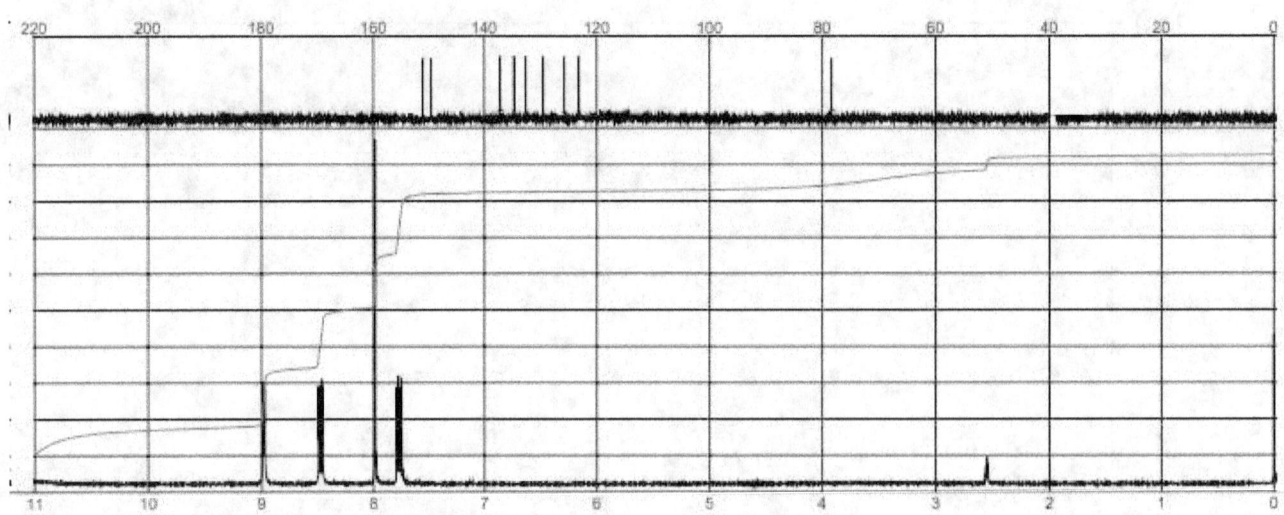

We will consider the ^{13}C nmr spectrum first.

^{13}C NMR Spectrum

Observations

Chemical shift δ (ppm)	Integral	Assignment
152	1	Aromatic ring*
150	1	Aromatic ring*
137	1	Aromatic ring
135	1	Aromatic ring
133	1	Aromatic ring
130	1	Aromatic ring
125	1	Aromatic ring
123	1	Aromatic ring
78	1	C – I

Conclusions

There is clearly a benzene ring which accounts for the singlets, all of integral one, those between δ 137 ppm and δ 123 ppm but there are eight aromatic carbon atoms. This implies that the molecule must contain two fused rings which share two carbon atoms. That, however, only accounts for eight carbon atoms and a pair of fused rings requires ten atoms in the aromatic system. If the molecule contains two fused rings but only eight aromatic carbon atoms, as well as the C – I bond, then the molecule must contain a nitrogen atom. The other peak must be due to a C – I bond and this carbon atom must also be somewhere in the fused rings.

There are two deshielded aromatic carbon atoms (marked with an *): one must be due to bonding to a nitrogen atom and the other due to bonding with an oxygen atom. The singlet at δ 78 ppm must be due to a C – I bond as iodine is not regarded as electronegative.

Given the number of aromatic carbon atoms and the relative lack of hydrogen atom it appears that the molecule comprises two fused aromatic rings. There are two possible base structures as shown below:

It is not possible to include oxygen in the base structure as it would require three bonds and that is not possible. It is possible to place the nitrogen atom in other positions in the right hand ring but rotation or reflection will turn them into one or other of the above structures.

There are two deshielded aromatic carbon atoms and since one will be bonded to the – OH substituent this prevents the right hand structure being feasible as there would be two deshielded carbon atoms due to the two bonded to the nitrogen atom. This means that the base structure must be:

and the three substituents can occupy one of two possible sets of positions:

It is reasonable to start with the assumption that the three substituents are placed on the, left hand, benzene ring

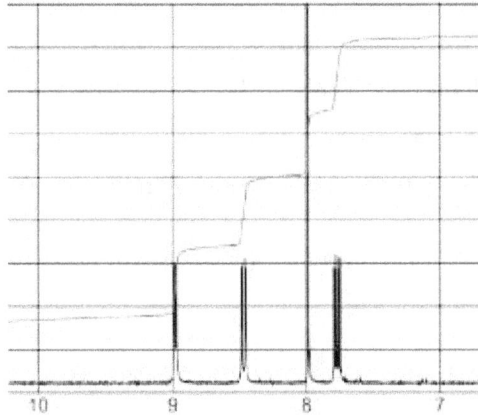

If neither of these work then we can consider a wider distribution of the three substituents.

It is clear that we cannot establish which is correct from the ^{13}C nmr so we now consider the ^{1}H nmr spectrum before consolidating the observations from the two spectra.

^{1}H NMR Spectrum

The expanded ^{1}H nmr spectrum is shown below:

This spectrum is a delight to analyse due to its simplicity.

There are three multiplets and one singlet all of integral one and all in the aromatic region. The signal at δ 2.6 ppm can be ignored as an impurity since the integral is far less than one.

To summarise the above spectrum we can observe:

- A triplet at δ 8.97 ppm;
- A doublet of doublets at δ 8.45 ppm;
- A singlet at δ 7.98 ppm;
- A doublet of doublets at δ 7.75 ppm.

The simplest possible structures are:

where R, R_1 and R_2 are Cl, I and OH for the following reasons:

- There will be a singlet due to the single hydrogen atom on the benzene ring.
- In all cases the hydrogen atoms on the pyridine ring will cause two doublets of doublets and a triplet:

The squiggles denote the, as yet to be determined positions, of the H, Cl, OH and I substituents but examining the pyridine ring, the signal will be the same for all three possible isomers, as follows:

- H_a will be split into a doublet by H_b and into a doublet of doublets by H_c;
- H_b will be split into a triplet by H_a and H_c;
- H_c will be split into a doublet by H_b and into a doublet of doublets by H_a.

Conclusions on the signals due to the pyridine ring:

- As there is only one triplet that must be caused by H_b , accounting for the triplet at δ 8.97 ppm.
- Due to its proximity to the aromatic, electronegative, nitrogen atom, H_c will be deshielded and appear at a higher chemical shift than would be the case if the heteroatom was not electronegative. This accounts for the doublet of doublets at δ 8.45 ppm.
- This leaves the doublet of doublets at δ 7.75 ppm which can be assigned to H_c.

Comfortable with the assignments on the pyridine ring we must now examine the structure of the benzene ring and we have three possible isomers:

where R, R_1 and R_2 are Cl, OH and I and it is the exact assignments that we need to determine.

The hydrogen atom will be a singlet in all three examples so can approach this problem in two ways:

- Consider the chemical shift of the singlet due to the sole hydrogen substituent on the benzene ring;
- Examine the ^{13}C nmr spectrum with particular attention to the benzene ring as the signals from the carbon atoms in the pyridine ring will be the same in all three isomers.

The chemical shift of the singlet is measured to exist at δ 7.98 ppm and this is significantly deshielded from the default δ 7.33.

To address this we must consider the electronegativities of the bonding atoms.

On the Pauling Scale of electronegativities, we have:

- Iodine: 2.7, so not classed as electronegative;
- Cl: 3.0;
- O: 3.5.

It is clear then that the significant bonding atoms are chlorine and oxygen and we need to consider the effect on the chemical shift of the sole hydrogen atom on the benzene ring.

This sole hydrogen atom will be more strongly deshielded by proximity to the chlorine substituent than by the oxygen atom so we can assume that, if we place chlorine at the top of the benzene ring then the hydrogen atom must be adjacent to it.

If the $-$ OH group is on the other side of the hydrogen atom then it would be even more deshielded which it is not and so this position must be occupied by the iodine atom, leaving the $-$ OH group para to the chlorine substituent.

This gives us the following structure,

which we can confirm or dismiss by examining, again the ^{13}C nmr spectrum.

If we relabel the molecule with fresh alphabetic labels we can attempt to assign the peaks:

Benzene (C_6H_6) produces a single resonance at δ 128.5 ppm and so any peaks *above this* are considered to be *deshielded* whilst resonant signals *below* δ 128.5 ppm are viewed as *shielded*.

On this basis, the spectrum of this molecule has six deshielded carbon atoms and three shielded carbon atoms. The three shielded carbon atoms are those furthest from any electronegative substituents or the integral nitrogen atom and could be assigned to C_g, C_h and C_i but this is a mistake since $C - I$ carbon atoms produce a chemical shift in the region δ 70 $-$ 90 ppm.

We can also assign the deshielded peaks to C_d and C_f as these are the two most electronegative substituents and they will will deshield the carbon atoms and shift the resonant signals significantly upfield.

There are five electronegative elements[1] and since nitrogen and oxygen both have an electronegativity of 3.0 on the Pauling scale and since the carbon signals are observed at δ 150 ppm and δ 152 ppm it is not possible to assign specific signals to each of the two carbon atoms.

This leaves us with six resonant signals to assign.

Cl is electronegative and will deshield C_a so we can assign the signal at δ 137 ppm to C_a and this leaves us with five signals to assign. These occur in the range δ 137 – 123 ppm.

The peak at δ 137 ppm can be assigned to C_e due to its proximity to the nitrogen atom and this accounts for the entire structure and confirms the structure to be:

We can make the following assignments based on the atoms' proximity to or distance from the electronegative atoms:

Chemical shift δ (ppm)	Integral	Assignment
152	1	C_d or C_f
150	1	C_d or C_f
137	1	C_e
135	1	C_a
133	1	C_b or C_i
130	1	C_b or C_i
125	1	C_g or C_h
123	1	C_g or C_h
78	1	C_c

Conclusions

Structure:

Systematic name: 5-chloro-7-iodoquinolin-8-ol

---

[1] Many writers and textbooks state that there are only four electronegative elements: N, O, F, Cl but since Br has an electronegativity close to that of Cl (2.8 and 3.0 respectively) and is used to explain electrophilic addition of HBr to a C=C bond we strongly believe that this should also be included as an electronegative element.

Chapter XIX

Carbamazepine

Discovered in 1953, ***carbamazepine*** has been used as an anticonvulsant medication since 1962 to treat epilepsy and neuropathic pain. In combination with a number of other medications it is used to treat bipolar disorder. There have been some studies suggesting efficacy for attention deficit and hyperactivity disorder (ADHD).

It has a number of minor side effects including dizziness, nausea, rashes and drowsiness but there are a number of serious side effects including impaired motor coordination, decreased bone marrow function, confusion and suicidal thoughts. It is not used for those with bone marrow issues nor for pregnant or breastfeeding women and consumption of alcohol is discouraged when the medicine is being taken.

It is a colourless crystalline solid with melting and boiling points of 190°C and 411°C.

Carbamazepine has elemental composition: C:76.18%, H:5.13%, N:11.86%, O:6.77% and a formula mass of 236.27 g mol^{-1} meaning that its **empirical** and **molecular** formulas are the same, $C_{15}H_{12}N_2O$.

Infrared Spectrum

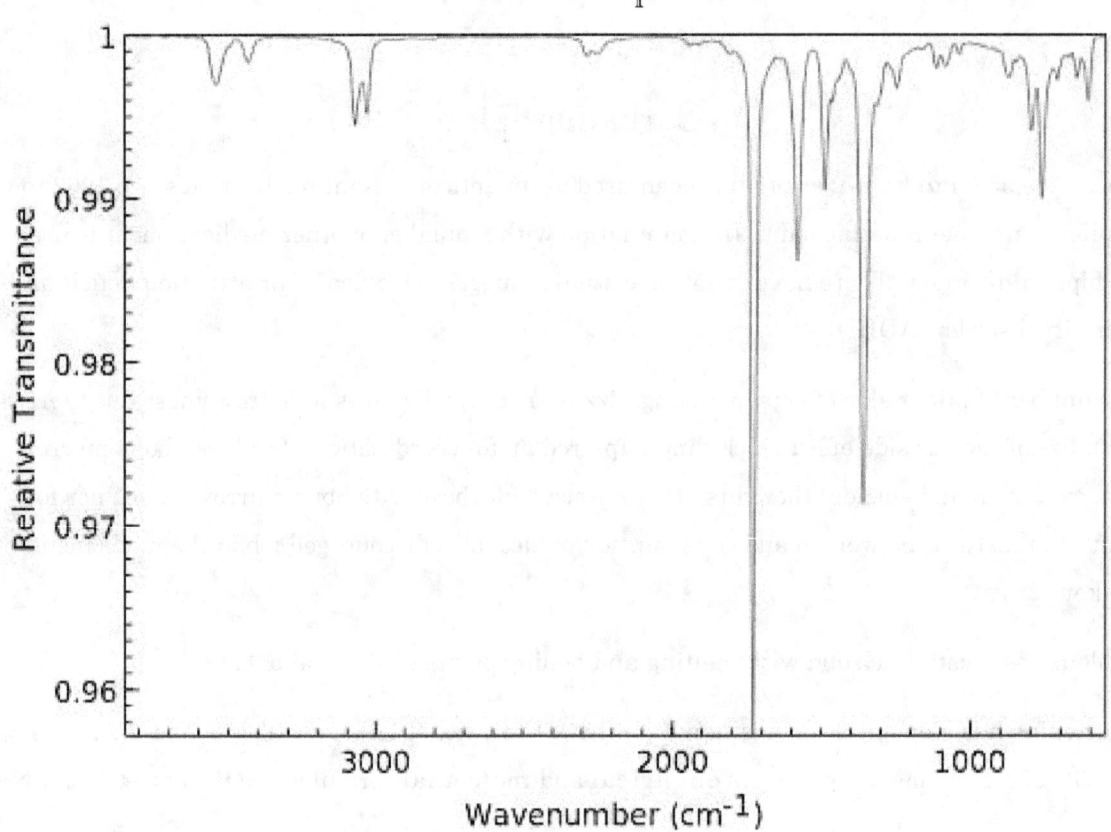

Observations

(√ / X)	Wavenumber range (cm⁻¹)	Wavenumber (cm⁻¹)	Assignment
X	3200 - 3700		O – H
√	3200 - 3600	3590, 3520	N – H
√	3000 – 3300	3080, 3010	C – H (aromatic)
X	2500 – 3000		C – H (aliphatic)
X	2200 – 2500		C ≡ N
√	1700 – 1800	1720	C = O
X	1600 – 1700		C = C (aliphatic)
X	1585 – 1600		C – C (aromatic)
X	1450 – 1600		C – C (aromatic)
X	1000 – 1300		C – O
X	700 – 1000		C – X (X = Cl, Br or I)

Conclusions

▪ This molecule is aromatic and contains both carbonyl, C = O, and amino, N – H, groups.

▪ The absence of peaks assignable to O – H or C – O groups indicates that the molecule is neither a carboxylic acid or an ester.

Mass Spectrum

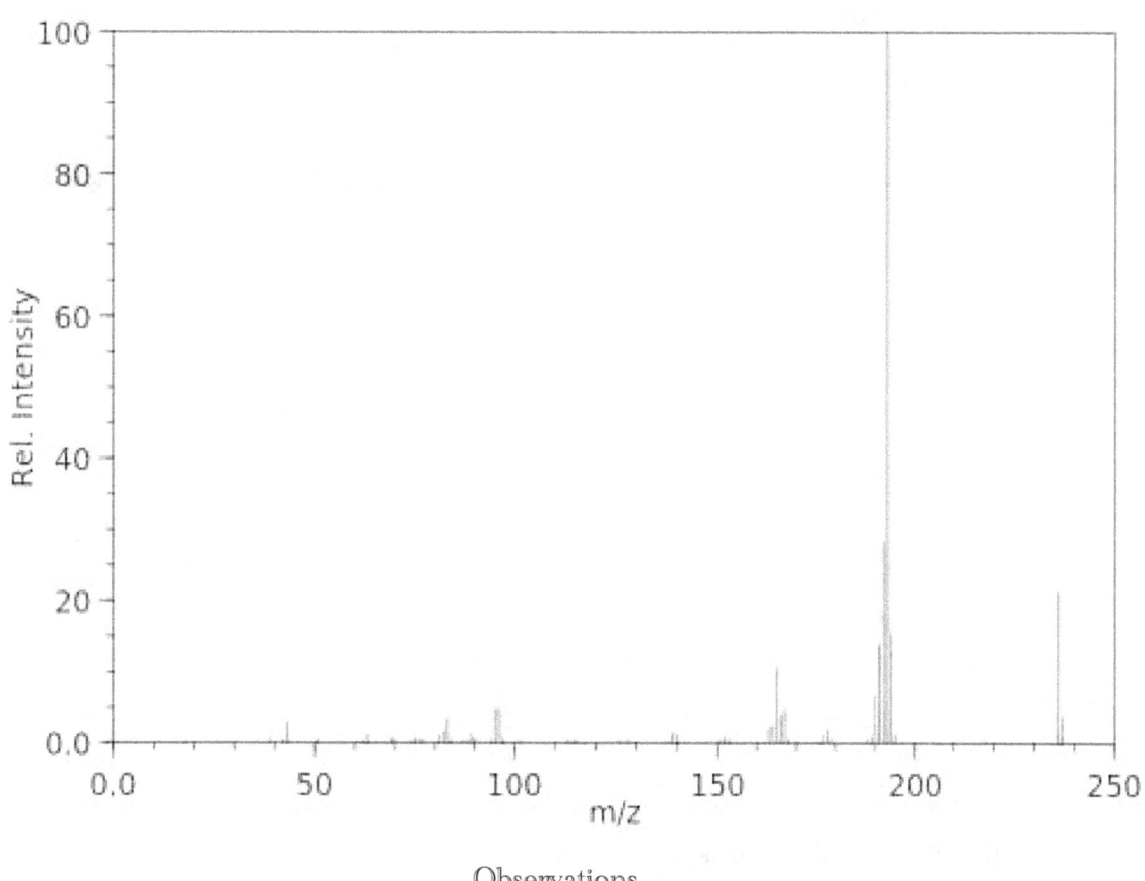

Observations

Charged fragments (m/z)	Assignment	Charged fragments (m/z)	Assignment
Molecular ion: 236	$[C_{15}H_{12}N_2O]^+$	**Base peak: 193**	$[C_{12}H_5N_2O]^+$

Charged fragments (m/z)	Assignment	Charged fragments (m/z)	Assignment
218	$[C_{15}H_{10}N_2]^+$	139	$[C_{11}H_7]^+$
178	$[C_{12}H_9N_2]^+$	96	$[C_7H_{12}]^+$
165	$[C_{12}H_7N]^+$	76	$[C_6H_4]^+$

Conclusions

There are insufficient hydrogen atoms for this molecule to be a linear molecule and so it must be aromatic. This conclusion is supported by the presence of the peak at m/z = 76 which can be assigned to one or more benzene rings with two substituents.

The presence of fifteen carbon atoms strongly suggests the presence of two or three fused rings or on ring with two other rings as substituents.

We can learn much more from the ^{1}H and ^{13}C nmr spectra which is our next task.

NMR Spectra

The ^{1}H and ^{13}C nmr spectra are displayed below:

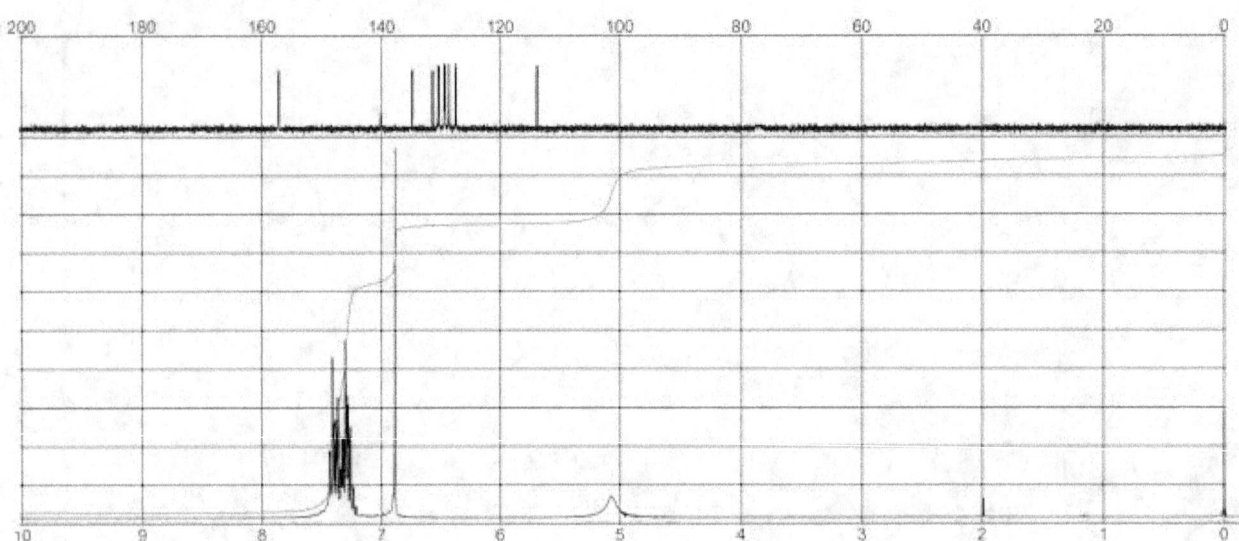

The ^{1}H nmr spectrum is extremely complex whilst the ^{13}C nmr spectrum is clear and so it makes sense to examine the latter first.

^{13}C NMR Spectrum

We can tabulate the ^{13}C NMR spectrum as shown below:

Chemical shift δ (ppm)	Integral	Assignment
165	1	C = O
135	2	Aromatic carbon
132	2	Aromatic carbon
131	2	Aromatic carbon
130	2	Aromatic carbon
129	2	Aromatic carbon
127	2	Aromatic carbon
116	2	C = C

There are twelve aromatic carbon atoms and they must be produced by two separate benzene rings. If present, the rings cannot be fused since they would then share two carbon atoms and the number would be lower than observed.

Since there is an alkene C = C bond one feasible, partial, structure is shown below:

where the structure above accounts for all the signals observed between δ 116 and δ 135 ppm.

The squiggle represents the remainder of the moleculewhich is of formula CH_2N_2O.

The fragment, CH_2N_2O, must contain an amide functional group, accounting for the presence of the carbonyl, $C = O$, functional group (indicated by the infrared spectrum). It cannot form the remainder of the ring as there will be nowhere to place the remaining H_2N_2 atoms.

This leaves us with only one option which is to place one of the nitrogen atoms in the ring and this could then allow us to add the amide group into the molecule as shown below:

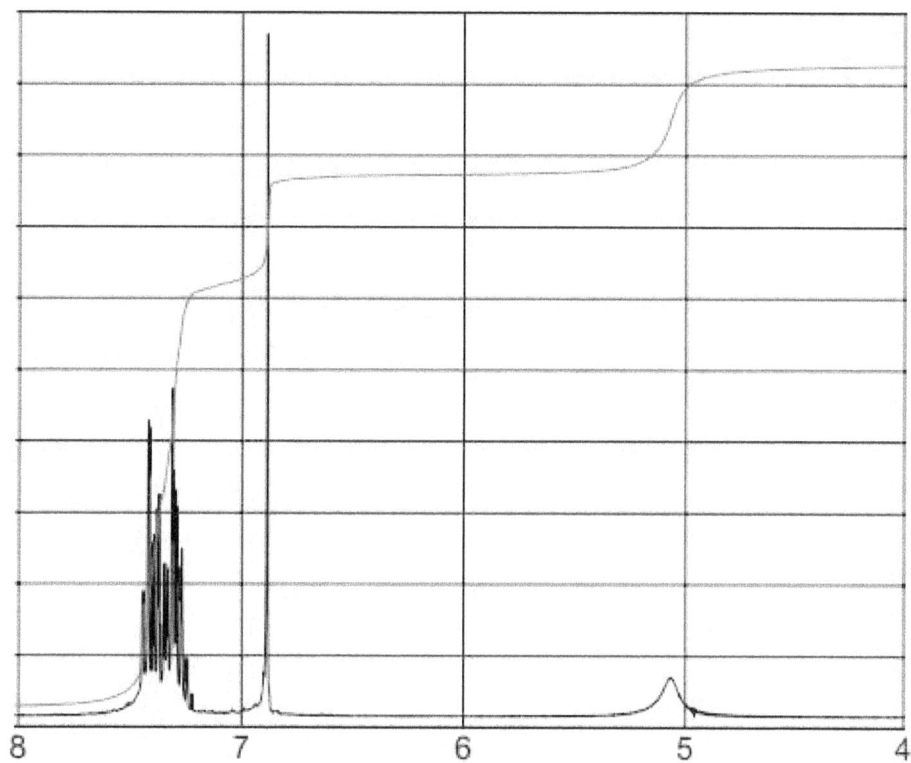

This accounts for all the atoms and also explains the presence of the N – H stretch in the infrared spectrum but we need to confirm the structure by examining the ^1H nmr spectrum which is our final task in this chapter.

To achieve this we need to label the hydrogen atoms as shown below:

^1H NMR Spectrum

The expanded ^1H nmr spectrum is shown below and it has a number of highly significant signals which are discussed below the spectrum.

■ The singlet, of integral two, at δ 6.9 ppm can be assigned to the alkene hydrogen atoms, H_a / $H_{a'}$. That it is a singlet is highly significant since it demonstrates that the functional groups on both sides of the alkene, C = C, double bond are identical since, otherwise, it would be a doublet and there would be a measurable coupling constant.

■ The broad peak at δ 5.1 ppm, of integral two, can be due to the N – H part of the amide group;

■ The complex multiplet, centred on δ 7.4 ppm, cannot be analysed as it is caused by a number of overlapping multiplets but the integrals comprise four sets all of of integral two which is consistent with the molecule containing four pairs of chemically and magnetically equivalent hydrogen atoms as shown previously.

■ Although we can draw the amide hydrogens, H_f, as shown above, in reality, the two hydrogen atoms are slightly different due to the planarity of the carbonyl group and this explains why the signal, of integral two, is not a sharp singlet.

Conclusions

Structure:

Systematic name: 5H-dibenzo[b,f]azepine-5-carboxamide

Chapter XX

Cyproheptadine

Cyproheptadine, sold under the brand name *Periactin* amongst some other others, is a first – generation antihistamine with additional local anaesthetic properties.

Patented in 1959, it was adopted for medical use in 1961 and is used to treat allergic reactions specifically hay fever but is also used as a preventative treatment for migraine. It is also used in cats as an appetite stimulant and in the treatment of asthma.

At room temperature this substance is a crystalline solid, with a melting point of 165°C and a boiling point of 440°C. It is often administered as a hydrochloride salt to make it soluble but the analysis is here for the organic molecule itself.

This compound has the elemental composition: C: 87.68%, H: 7.38%, N: 4.87% and a formula mass of 287.41 g·mol^{-1}.

This means that the **empirical** and **molecular** formulas are both $C_{21}H_{21}N$.

Infrared Spectrum

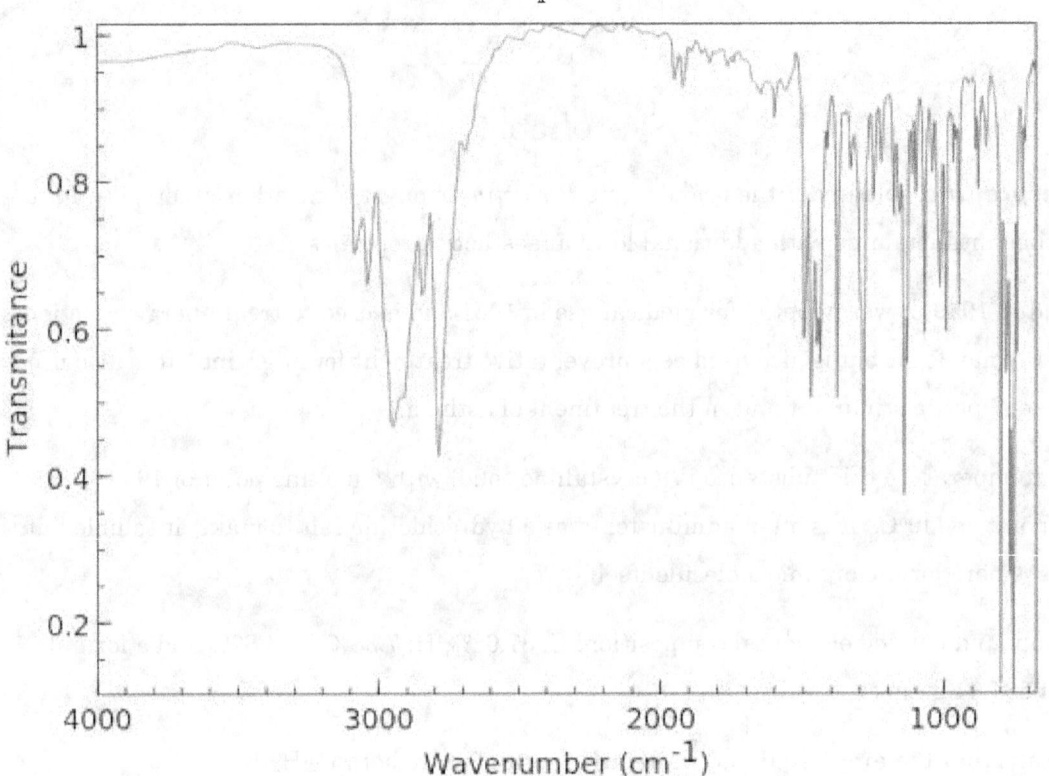

Observations

(√ / X)	Wavenumber range (cm⁻¹)	Wavenumber (cm⁻¹)	Assignment
X	3200 - 3700		O – H
X	3200 - 3600		N – H
√	3000 – 3300	3130,3100	C – H (aromatic)
√	2500 – 3000	2995,2870,2780	C – H (aliphatic)
X	2200 – 2500		C ≡ N
X	1700 – 1800		C = O
X	1600 – 1700		C = C (aliphatic)
X	1585 – 1600		C – C (aromatic)
√	1450 – 1600	1530, 1510	C – C (aromatic)
√	1000 – 1300	1290	C – O or C – N
X	700 – 1000		C – X (X = Cl, Br or I)

Conclusions

- From the peaks above 3000 cm⁻¹, this compound must be aromatic;

- The peaks to the right of 3000 cm⁻¹ indicate that is has one or more aliphatic groups;

- The peak at 1290 cm⁻¹ indicates the presence of a C – N bond.

- The absence of any peaks in the 3200 – 3600 cm⁻¹ shows that there is no N – H bond. This means that the nitrogen atom must be substituted with alkyl groups or exist in a ring.

Mass Spectrum

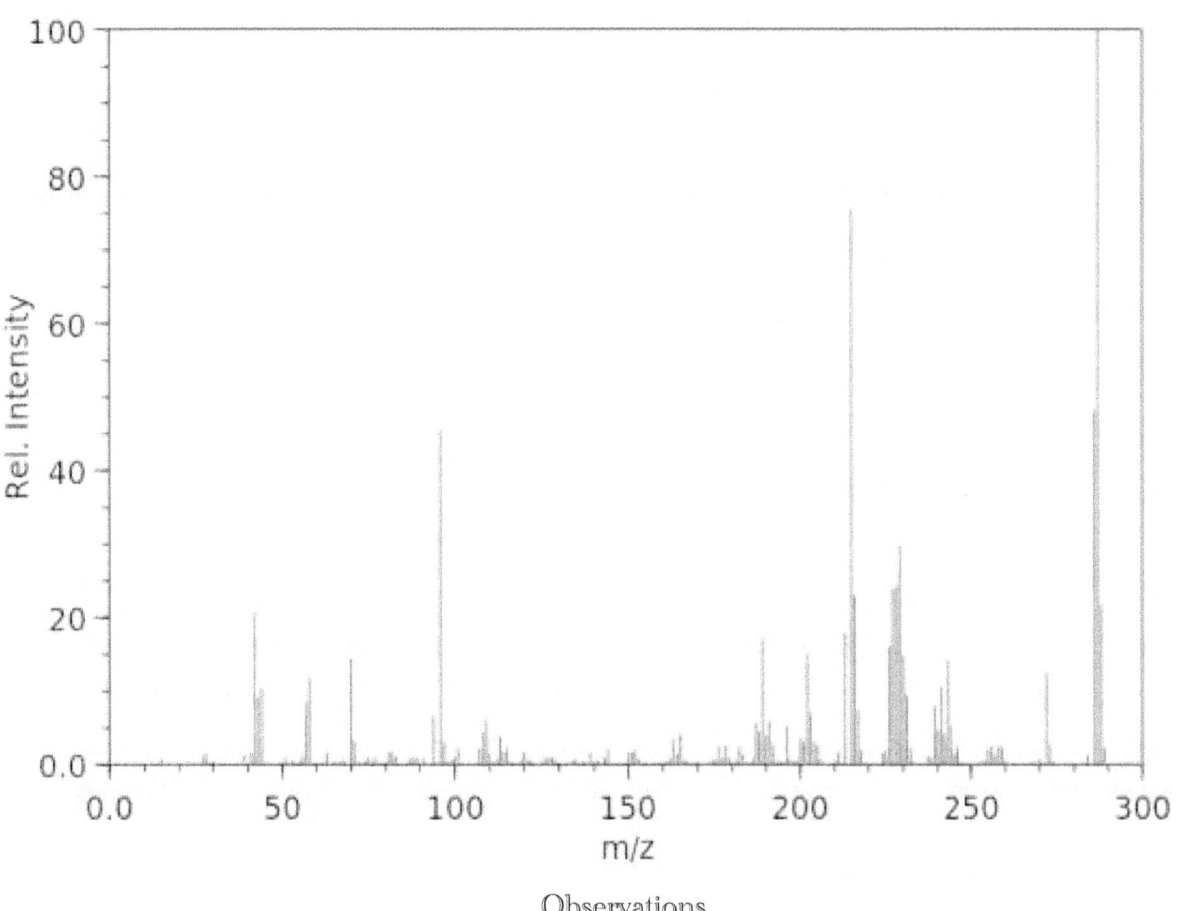

Observations

Charged fragments (m/z)	Assignment	Charged fragments (m/z)	Assignment
Molecular ion: 287	$[C_{21}H_{21}N]^+$	Base peak: 287	$[C_{21}H_{21}N]^+$

Charged fragments (m/z)	Assignment	Charged fragments (m/z)	Assignment
272	$[C_{21}H_{20}]^+$	189	$[C_{15}H_9]^+$
229	$[C_{18}H_{13}]^+$	96	$[C_7H_{13}]^+$
215	$[C_{17}H_{11}]^+$	70	$[C_5H_{10}]^+$
202	$[C_{16}H_{10}]^+$	42	$[C_3H_6]^+$

Conclusions

It is difficult to conclude much from this spectrum as the assignments can appear to be due to a fragmenting long chain or a shorter, highly substituted molecule. An alternative is that there are two or more fused rings, whether aromatic or aliphatic. This makes sense since the molecular formula is $C_{21}H_{21}N$ and, given a limited number of hydrogen atoms, it is difficult if not impossible to draw a structure which would be a long chain or shorter, but substituted chain.

We can learn more by examining the ^1H and ^{13}C NMR spectra which is our next task.

NMR Spectra

The ^1H and ^{13}C NMR spectra are displayed below.

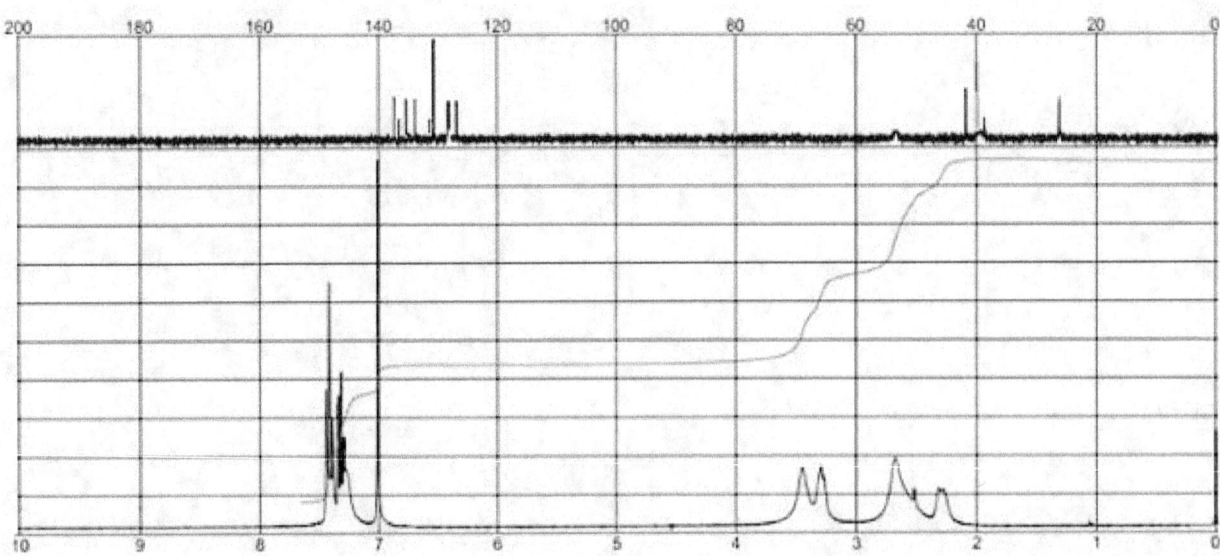

Initially, the ^1H nmr spectrum appears extremely difficult to analyse although, due to the cumulative integral, it does appear that the molecule is substantially aromatic with more than one aromatic ring. The broad peaks, δ 2 – 3 ppm indicate the presence of both a CHx group attached to a benzene ring and also the presence of a HC – N group whilst the δ 3 – 4 ppm signals imply the presence of both an N – H group and alkene hydrogens.

We will consider the ^{13}C nmr spectrum first.

^{13}C NMR Spectrum

The ^{13}C nmr spectrum has three distinct sets of signals: δ 120 – 138 ppm (aromatic carbon atoms), δ 29 – 42 ppm (C – N group) and δ 24 ppm (C – C group).

In detail, the ^{13}C nmr spectrum contains the following peaks to which we can add tentative assignments:

Chemical shift δ (ppm)	Integral	Assignment
130	4	Aromatic
134	2	Aromatic
136	2	Aromatic
137	1	Aromatic
138	2	Aromatic
26	2	Aromatic
42	2	C – N or C – C
38	1	C – N or C – C
26	2	C – N or C – C

The peaks at δ 42, 38 and 26 ppm cannot immediately be assigned just from their chemical shifts as the C – N and C – C regions overlap (as can be seen from the data sheet). Since however, there is only one nitrogen atom in the molecule it is immediately apparent that the peak as δ 38 ppm can be assigned to a C – N bond as that is the only peak of integral one.

Jumping back to the infrared spectrum, there is no evidence for an N – H bond and so the nitrogen atom must be bonded to a terminal alkyl group or to another carbon atom in the rest of the molecule. Noting the valencies of the atoms, there are two other peaks on the right of the spectrum to assign, those at δ 42 and δ 26 ppm. Both of these are of integral two and this leads to the suggestion that the molecule contains the following grouping:

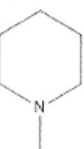

This ring must be bonded to the rest of the molecule so we can envisage that it is bonded to one of the carbon atoms in the ring or to the nitrogen atom. We can, however, disregard the suggestion that the nitrogen atom bonds to the rest of the molecule as that would too few hydrogen atoms to assign to the rest of the molecule. There is also no evidence from the ir spectrum for the presence of an N – H bond so that atom must be bonded to a methyl group.

This leaves us with two possible types of bonding of this group to the rest of the molecule:

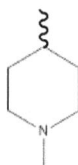

where the squiggle can be a C – C or a C = C bond.

The formula mass of this fragment is $C_6H_{11}N$ which leaves us with the rest of the formula mass, $C_{15}H_{10}$, to be assigned. There is no possible fragment which can be assigned to this formula and so the bonding must be a C = C bond which gives us the following functional grouping:

If this is correct then there should be a C=C peak in the ir spectrum (~1650 cm^{-1}). There are scrappy little peaks there but they might be obscured by the very strong peaks observed to the right of 800 cm^{-1} so we can neither accept or deny this possibility. There are, however, strong peaks just to the right of 1500 cm^{-1} so it is possible that one of them might be due to a C = C bond.

Returning to the functional group, the formula mass of the fragment will be 97 g mol^{-1}. Now, jumping back to the mass spectrum, there is a tiny peak at that value and a strong peak at m/z = 96 of relative intensity 52% which could be assigned to a fragment of formula $[C_6H_{10}N]^+$ This peak is the third highest in the mass spectrum and so is extremely important.

This means that we can, at this point, neither accept nor deny this possibility but it is worth progressing as there is no definitive evidence for its non-existence although the ^{13}C nmr spectrum is very compelling.

If we accept this possibility then we have to assign a structure to the remaining fragment, taking into account the ^{13}C and ^{1}H nmr spectra. We might also find supporting evidence from the infrared and mass spectra but we can focus on the ^{13}C nmr spectrum first.

From the ^{13}C nmr spectrum, we can see that the molecule contains fifteen fused, aromatic, carbon atoms. Subtracting the previously accounted for hydrogen atoms (from the previously suggested substituent group), this gives us a fragment of formula $C_{15}H_{10}$.

Once we have discounted the possibility of a long chain hydrocarbon or a shorter, highly substituted chain we are left with only the possibility of aromatic groups. This is confirmed by the ^{13}C nmr spectrum which shows a cluster of aromatic carbon resonances which must be assigned. It is not a bad starting point to assume that if there are at least twelve aromatic carbon atoms then we are dealing with at least two benzene rings.

That will account for twelve of the carbon atoms and a significant number of the remaining hydrogen atoms, given that we have already accounted for the sole nitrogen atom.

This means that we can envisage a structure of the form below (where the squiggles describe indeterminate bonds):

If we accept that the lower two squiggles represent $- C - C -$ bonds then we have the following structure

We can analyse the signals in the ^{13}C NMR spectrum in two separate groups: aliphatic and aromatic. We will start with the resonance signals in the aliphatic region but before we start it is important to remember that all peaks in a ^{13}C nmr spectrum are singlets and we do not observe the multiplicities apparent in ^{1}H nmr spectra.

There is no mystery about this as it is simply because ^{13}C constitutes only about 1.3% of all carbon atoms. The vast majority are ^{12}C isotopes which, with an even number of nucleons (protons and neutrons), do not resonate. It is highly unlikely that, even with the largest molecules, there will be more than a handful of ^{13}C isotopes present and the likelihood of them being bonded to each other is vanishingly small. If they were then there would be the multiplicities observed in ^{1}H nmr spectra.

XX - Cyproheptadine

It is proposed that this molecule has the structure which comprises both aromatic and aliphatic regions:

Mainly aromatic ?

Aliphatic

We will start by considering the aliphatic portion of the molecule.

Aliphatic resonance peaks

For the lower, aliphatic, ring we will have the following:

1. Two pairs of chemically and magnetically equivalent carbon atoms as shown below:

2. A singlet due to the methyl, $- CH_3$, group:

This accounts for the three peaks in the aliphatic region, δ 28, 36 and 42 ppm of integrals 2, 1, 2 respectively but we now have to assign the three peaks using, as a guide, the data sheet at the beginning of this volume.

A. We can assign the peak, δ 36 ppm of integral one, to the $N - CH_3$ carbon atom.

B. We have two pairs of carbon atoms which are chemically and magnetically equivalent within the pair but the two pairs are not chemically and magnetically equivalent between the pairs it is not possible to assign individual chemical shifts until we consider that two are bonded to the electronegative nitrogen atom which deshields the two carbon atoms.

   The two carbon atoms bonded to the nitrogen atom are deshielded due to the electronegativity of nitrogen and will be of higher chemical.. This is the peak at δ 42 ppm.

C. This leaves us with the peak of integral two due to the other two carbon atoms which are mutually chemically and magnetically equivalent.

That the methyl group is bonded to the electronegative nitrogen atom means that the methyl carbon atom is deshielded and explains why it appears as the middle aliphatic carbon signal.

It is easier to examine this visually and so we have the following:

δ 28 ppm   δ 28 ppm

δ 42 ppm   δ 42 ppm

δ 36 ppm

The unlabelled carbon atom at the top of the fragment is discussed next before we investigate the signals in the aromatic region of the ^{13}C nmr spectrum.

D. There is a singlet due to the single carbon at the top of the aliphatic ring:

but the question is **Does it count as aliphatic or aromatic**?

We can only decide that once we fit in the remaining atoms i.e. replace the squiggle at the top of the molecule.

The molecular formula is $C_{21}H_{21}N$ and we have accounted for the heterocyclic ring and substituent at the bottom. Given the ratio of hydrogen to carbon atoms and the number of peaks in the aromatic region this molecule must have at least two aromatic rings since nothing else fits. In principle, it would be possible to have the rings fused as in naphthalene but there is no way in which such a molecule could be constructed.

If we assume that the structure discussed is correct then we have accounted for all but two carbon and two hydrogen atoms and so, if we continue with this model, we must replace the squiggle with C_2H_2 which means it must be an $- HC=CH -$ group and so we can draw the molecule as:-

We are now in a position to examine the aromatic region of the ^{13}C nmr spectrum (the ^1H nmr spectrum is not much use at this stage).

The aromatic region contains the following signals:

Chemical shift δ (ppm)	Integral	Assignment
130	4	Aromatic
134	2	Aromatic
136	2	Aromatic
137	1	Aromatic
138	2	Aromatic
128	2	Aromatic
127	1	Aromatic
126	2	Aromatic

The assumption that the molecule is aromatic, due to the lack of sufficient hydrogen atoms to be aliphatic, is confirmed by both the infrared spectrum and the ^{13}C nmr spectrum. If we look at this structure in more detail we can see that there are certain symmetries:

1. The highlighted carbon atoms account for the peak of integral of four (δ 130 ppm)

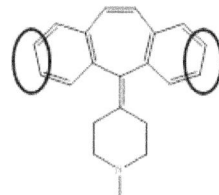

2. Three peaks of integral two appear are caused by the carbon atoms indicated below (the chemically and magnetically equivalent carbon atoms are highlighted with matching shapes):

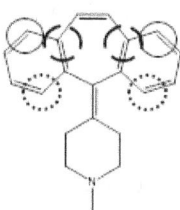

3. A peak of integral one is due to the atom as shown below:

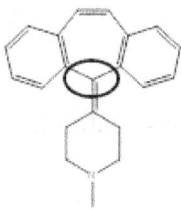

4. This accounts for all but one peak of integral two.

   This can only fit where the squiggly line was and must be a symmetrical alkene, – C = C –, group, highlighted by the rectangle.

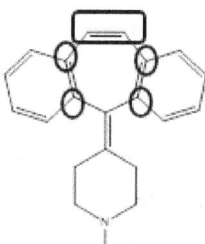

This alkene grouping, highlighted by the rectangle, is not part of the aromatic system.

Aromatic groups must have hydrogen atoms on all their parent atoms in order to enable delocalisation of the electrons and if we examine this structure we can observe that the carbon atoms immediately to either side of the highlighted alkene and on the other side of the ring do not have hydrogen atoms (highlighted by circles).

This structure perfectly matches the ^{13}C nmr spectrum so we now only need to have another look at the ^{1}H nmr spectrum which is nowhere near as useful due to massive overlap of the peaks in the aromatic region and the broad peaks in the aliphatic region.

If we examine the proposed structure again we can observe that it contains ten aromatic or alkene hydrogen atoms and eleven aliphatic hydrogen atoms. The ratio of the integrals in the matches the ratio of aromatic/alkene : aliphatic hydrogen atoms of 10 : 11 but the most significant signal is that at δ 7 ppm which accords with the existence of the alkene group that we have just discussed.

Conclusions

Structure:

Systematic name: 4-(5H-Dibenzo[a,d]cyclohepten-5-ylidene)-1-methylpiperidine

Other trivial names: Periactin

Notes:

Systematic, IUPAC, names are essential to ensure that different chemists are talking and writing about the same molecule but the names can be horrendously complicated. Nevertheless the name does require some explanation.

This molecule is an example of a *benzocycloheptene*.

Cycloheptenes are seven – membered carbon rings containing a C = C bond as shown and the basic molecule of this class, cycloheptene, is shown below:

Benzocycloheptenes contain one or more benzene rings bonded to the cycloheptene ring and this explains why the systematic, IUPAC, name contains the term although it becomes more complicated with the numbering of the C = C bond.

The final part of the name, *1 – methylpiperidine*, however originates from the name of the compound shown below which acts a functional group to this molecule.

www.ingramcontent.com/pod-product-compliance
Lightning Source LLC
Chambersburg PA
CBHW081056170526
45166CB00006B/2080